Using Social Science to Understand and Improve Wildland Fire Organizations:

An Annotated Reading List

Authors:

Gregory Larson, Associate Professor
Communication Studies, University of Montana
Missoula, MT

Vita Wright, Research Application Program Leader
Aldo Leopold Wilderness Research Institute, Missoula, MT
Rocky Mountain Research Station
U.S. Department of Agriculture, Forest Service

Cade Spaulding, Graduate Student
Communication Studies, University of Montana
Missoula, MT

Kelly Rossetto, Graduate Student
Communication Studies, University of Montana
Missoula, MT

Georgi Rausch, Graduate Student
Communication Studies, University of Montana
Missoula, MT

Andrea Richards, Graduate Student
Communication Studies, University of Montana
Missoula, MT

Stephanie Durnford, Graduate Student
Communication Studies, University of Montana
Missoula, MT

United States Department of Agriculture / Forest Service

Rocky Mountain Research Station

General Technical Report RMRS-GTR-201
September 2007

Acknowledgments

We thank Jim Saveland for conceiving of, and securing funding for, this project. We also appreciate Dave Thomas' support from project initiation through completion, including his assistance in annotating approximately 20 books and articles. Both Jim and Dave suggested many readings that greatly improved the content of this list. Karl Weick, Dave Thomas, Mike DeGrosky, and Linda Langner provided in-depth reviews, ideas on organization, and suggested readings. We received additional helpful suggestions from the following individuals: Paul Chamberlin, Jim Cook, Jon Driessen, Don MacGregor, Dick Mangan, Jerry Pepper, Ted Putnam, and Larry Sutton. This project was funded by the USDA Forest Service National Fire Plan, USDA Forest Service Rocky Mountain Research Station, The University of Montana, and the Wildland Fire Lessons Learned Center.

Author Bios

Gregory Larson is Associate Professor in the University of Montana's Department of Communication Studies. Greg obtained his Ph.D. at the University of Colorado in 2000; his research emphases include: organizational culture, communication technologies, and occupational identity. Greg has published in journals such as Communication Monographs, The Journal of Applied Communication Research, Management Communication Quarterly and Organizational Dynamics. He became interested in the management of wildland firefighting in 1994, while living in Colorado during the aftermath of the South Canyon fire.

Vita Wright is Research Application Program Leader at the Aldo Leopold Wilderness Research Institute, Rocky Mountain Research Station, USDA Forest Service. She obtained her Master's degree in ecology in 1996, and has worked as a research application specialist since 1998. Currently working on her Ph.D. through the University of Montana, College of Forestry and Conservation, she is studying personal and organizational influences to the use of science by federal agency fire managers.

Cade Spaulding, while authoring this publication, was a graduate student in communication studies at the University of Montana, Missoula. He obtained his Master's degree in 2005 while studying the identity and identification among the Missoula Smokejumpers. He is currently working on his Ph.D. in communication at Texas A&M University, College Station.

Kelly Rossetto, while authoring this publication, was a graduate student in communication studies at the University of Montana, Missoula. She obtained her Master's degree in 2005 while studying parental comforting strategies, goals, and outcomes following the death of a child. She is currently working on her Ph.D. in communication at the University of Texas, Austin.

Georgi Rausch, while authoring this publication, was a graduate student in communication studies at the University of Montana, Missoula. She obtained her Master's degree in 2005 while studying identity in nonprofit arts organizations. She is currently working on her Ph.D. in communication at the University of Utah, Salt Lake City.

Andrea Richards, while authoring this publication, was a graduate student in communication studies at the University of Montana, Missoula. She obtained her Master's degree in 2006 while studying verbal negotiation in romantic relationships. She is currently working on her Ph.D. in communication at the University of Texas, Austin.

Stephanie Durnford obtained her Master's degree in communication studies from the University of Montana in 2005.

Contents

Preface

Reading as a Practice in Mindfulness

The world of fire management and leadership is becoming increasingly complex. There are greater demands for improved safety performance, reduced suppression costs, increased fuel treatment targets, no escapes, and no mistakes. Plates are overflowing, while the fire manager's time and attention are scarce resources in high demand.

Why then should a wildland firefighter want to take precious time to read one of the papers or books cited in this bibliography? On the surface, these published social science documents might seem to be written for academics and government researchers, holding little interest for on-the-ground fire managers.

So, the question—"Why take the time to read a social science document?" is fair, and we answer our own question with one word—*mindfulness*. To take the time to closely read a few of these books and articles that dwell on the human side of fire management (and to discuss them with a workmate) is to invite one to be more alert to the subtleties and nuances of the fire environment, where a mixture of weather, fuels, topography, and heat can combine with the most complex ingredient of all—the human being.

In a recent paper, organizational psychologist, Karl Weick and firefighting human factors psychologist and researcher, Ted Putnam, discuss the distinctions between eastern *mindfulness,* western *mindfulness*, and western *mindlessness* (Weick and Putnam 2006). To be *mindful,* from a western perspective, they note, is to seek out novel distinctions and to be careful of a thinking process that fixates on single perspectives without awareness that "things could be otherwise".

To be *mindless,* from the western perspective, is to make sense of the world using old recipes, standard operating procedures, and routines that are often enacted on auto-pilot. Mindlessness increases the wildland fire manager's chances of falling prey to serious unexpected events while he/she works within the always dangerous fire environment.

Using the western definition provided by Weick and Putnam, we can make the case that reading is an acute act of mindfulness as useful to the firefighter as after action reviews, simulations, sand table exercises, tactical decision games, staff rides, or all the other assorted activities associated with becoming a better firefighter. Reading about new concepts can disrupt our routines and cause us to view the world in a totally different light; reading increases requisite variety and makes the work world more complex and nuanced, so there is little doubt that reading a few of the articles and books cited in this annotated bibliography will increase one's mindfulness as a fire manager.

High Reliability Organizing (HROs) uses the word "mindfulness". Navy crewmen on aircraft carriers talk about "having the bubble"—the big picture of current operations, and as firefighters, we talk about the importance of "situational awareness" to firefighter safety.

To have these capacities in high-tempo, high stress situations, we need to practice *reflection skills* as well as *action skills* to become better at "reflecting-in-action", for "reflection-in-action" is closely related to "mindfulness", "having the bubble", and "situational awareness". We practice action with simulations, tactical decision games, and sand table exercises. We practice reflection with after action reviews and staff rides, journaling, and reflective conversation ("dialogue"). Combining reflection and action is what a learning organization is all about.

Reading can help us become more reflective. Ron Heifetz, one of the leading scholars on leadership, uses the metaphor of moving back and forth between the dance floor (where the action is) and getting on the balcony (where we can see and reflect on the big picture) (Heifetz and Laurie 1997). Taking a moment to read one of these articles is a form of getting on that "balcony" and pausing a moment to view the "dance floor" from a different angle with a fresh perspective.

Through reading articles and books such as those included in this reading list, one will enter what Mary Catherine Bateson (Bateson 2004) calls "the fellowship of lifelong learners", or, to paraphrase the words of former Zig Zag Hotshot Superintendent Paul Gleason, one will take a giant step toward becoming a lifelong "student of [the human side of] fire".

The articles and books listed here are a distillation of hundreds of possible entries that could have been included. They were selected by students, professors, on the ground fire practitioners, and federal researchers as excellent jumping off points for fire managers who want to become more knowledgeable about fire and the social sciences and more mindful about how human beings interconnect to make sense of the fire environment.

Our philosophy of reading—why professionals in all walks of forest fire management can sharpen their leadership abilities through reading—parallels the "Professional Reading Program" described by the Wildland Fire Leadership Program at the National Interagency Fire Center in Boise, Idaho: "*This [reading] is not busy work; this is not drudgery. These readings will provoke reflection, discussion, and debate. The selected titles have been chosen for their intrinsic excitement as well as their content. Many of the books will be hard to put down. Let this be your roadmap to an enjoyable and rewarding reading program*" (Wildland Fire Leadership Development Program 2005).

Here's to a successful lifetime of reading with the goal of becoming a more mindful fire manager.

Dr. Jim Saveland
*duBois Center for Individual
& Organizational Effectiveness
Rocky Mountain Research Station
Fort Collins, Colorado*

Dave Thomas
*Former member NWCG
Social Science Task Group
Renoveling
Ogden, Utah*

References

Bateson, Mary Catherine. 2004. **Willing to learn: passages of personal discovery.** Steerforth Press. Hanover, New Hampshire. 384 p.

Heifetz, Ronald A.; Laurie, Donald L. 1997. **The work of leadership.** Harvard Business Review. January: 124-134.

Wildland Fire Leadership Development Program. 2005. **Wildland fire book on books.** National Wildfire Coordinating Group. Boise, ID. Available: http://www.fireleadership.gov/toolbox/documents/pro_reading_room.htm.

Weick, Karl; Putnam, Ted. 2006. **Organizing for mindfulness: Eastern wisdom and Western knowledge.** Journal of Management Inquiry 15(3): 275-287.

INTRODUCTION

The wildland fire community has spent the past decade trying to understand and account for the role of human factors in wildland fire organizations. The first Wildland Firefighters Human Factors Workshop marked an important step in bringing social scientists and firefighters together to work on common problems related to firefighter safety. Efforts to understand how human factors apply to fire management and firefighter safety have been documented in the 1995 Wildland Firefighters Human Factors Workshop proceedings (Putnam 1995); Tridata studies (1996a, 1996b, 1998); Karl Weick's assessment of the Mann Gulch (Weick 1993) and Dude (Weick 2002) fires; the first Managing the Unexpected Workshop (Keller 2004); and the 2005 Wildland Firefighter Safety Summit proceedings (Butler and Alexander 2005).

Each of these efforts recognized that firefighters and fire managers face a broad array of mental, communication and management challenges. We hope firefighters and managers will find additional insight into how to address some of these challenges by accessing some the readings annotated in this list. The reading list is based on input from members of the fire community, the "human factors" work that has been done since South Canyon, and management-oriented literature on the social aspects of communication.

A significant amount of social science research relevant to the management of firefighters and firefighter safety has been conducted in disciplines such as psychology, sociology, management, and communication. This literature has been published primarily for scientific and business audiences, and much of the firefighting community has not seen it. Here, we have compiled and organized knowledge from the social sciences so that it can be used to manage organizational culture and practices related to firefighter and public safety, assess the effectiveness of firefighter safety campaigns, and improve firefighter safety trainings. More specifically, we have compiled literature on decision making and sensemaking, organizational culture and identity, leadership and change, organizational learning, and team and crew dynamics that may be helpful for understanding influences to individual behavior within an organizational setting. We have grouped the readings into topics so that readers can gain an initial familiarity with pertinent concepts through topical section summaries, and then add depth to existing knowledge through specific article annotations.

A key challenge within the fire management community is making this type of knowledge available for use at all levels in the fire organization. Through this reading list, we have attempted to increase access to, and understanding of, concepts that will be helpful to safely meeting fire management objectives. The next step, and a more challenging one, will be for readers to draw lessons from this literature that can help them prepare for and manage problems they encounter. In the shadow of past successes and tragedies, we acknowledge the difficult and essential work fire mangers do on a daily basis. We hope this will be a useful resource to the fire management and firefighting community.

Scope and Organization

The references included in this volume represent a diverse collection of classic and contemporary social science research related to managing organizations. Although not comprehensive, this collection provides a starting point to explore key issues related to management. Readers will primarily find two types of readings: 1) books, written with practitioners in mind, which tend to summarize and tie together research programs (such as Weick and Sutcliffe's "Managing the Unexpected") and 2) academic journal articles that report original research. These journal articles tend to focus more on specific issues than the books. They also form the foundations for organizational science. Since the goal of this volume is to increase awareness of potentially useful social science on organizational management, it does not include well known and highly recommended fire books such as Norman Maclean's "Young Men and Fire" or and Stephen Pyne's "Fire in America".

The sources selected for inclusion provide insight into key organizational research that will, hopefully, help firefighters and fire managers figure out new methods of dealing with both routine aspects of their everyday work life and the rarer and more complex fire situations in which they often find themselves.

The bibliography that follows is organized into three main sections:

- Human Factors and Firefighting
- Foundations for Understanding Organizations
- Understanding Organizations in High Risk Contexts

The first section, *Human Factors and Firefighting*, makes the case that "human factors" research, and social science research in general, are important tools for developing a complete understanding of firefighter safety. This section introduces readers to key work done in this area since the South Canyon Fire in 1994.

The second section, *Foundations for Understanding Organizations*, focuses on research that provides a solid foundation for understanding organizational dynamics. In this section, we focus on key topics including decision making and sensemaking, organizational culture, identification and identity, leadership and change, organizational learning, and teams and crews.

The third section, *Understanding Organizations in High Risk Contexts*, explores issues related directly to organizations that deal regularly with risk, uncertainty, and crisis. This section includes sub-sections on risk/uncertainty, high reliability organizing and crisis communication.

The second and third major sections are divided into topics. In each major section, and for each topic, there is a brief summary of the literature in that section, followed by an annotated list of readings. The annotations highlight relevant points and enable readers to determine whether the specific article or book is likely to be helpful for meeting their reading goals. While annotations were written to reflect the original authors' content, we have often used the last sentence to comment on the relevance of the article to this reading list.

The reading list concludes with a section that reviews some of the Internet resources available for readers who are interested in fire safety, leadership, and communication.

Intended Audiences

This annotated reading list is intended to serve those in the fire community who are interested in learning more about how the social sciences can be used to identify and solve diverse fire management and communication problems. This includes fire agency managers, trainers, incident commanders, firefighters, line officers, and others working to assess or improve firefighter safety. For those who have read widely in the organizational literature, we hope this volume will provide additional resources of interest. Such readers may find the added depth and understanding they seek by reading journal articles that report on original research findings. For those readers without a broad background in the management research literature, this volume provides many resources with practical applications for your use. Such readers may want to pay particular attention to the suggested readings listed in Appendix A and shown as "Author's Picks" throughout the document.

For ease of use, we:

- Divided the readings into the three main sections previously listed.
- Summarized each of the main sections and topic areas.
- In Appendix A, identified a scaled-down list of books/ articles that readers can use to quickly gain familiarity in each topic area. These readings are shown as "Author's Picks" throughout the document.

Obtaining These Readings

We recognize that it may be difficult for those in the field to obtain many of the readings included in this publication.

Where electronic access is possible, we have included on-line addresses in the citations and/or posted articles to the Wildland Fire Lessons Learned Center Library (http://www.wildfirelessons.net/Library.aspx). For those with access to local public or University libraries, it is also possible to obtain books and articles through interlibrary loan programs. For those who work in more remote locations, there are several government resources that provide document delivery (free to Federal employees in the United States). A current list of these resources can be found on the Aldo Leopold Wilderness Research Institute's website (http://leopold.wilderness.net/library.htm). They include:

- Bureau of Land Management Library (http://www.blm.gov/nstc/library/library.html)
- National Agricultural Library (http://www.nal.usda.gov/services/request.shtml)
- National Forest Service Library (http://fsweb.wo.fs.fed.us/library)
- National Park Service Library (http://library.nps.gov)
- USDA Digital Desktop Library (http://www.nal.usda.gov/digitop)
- USDA FS Research Publications (http://www.treesearch.fs.fed.us)
- USDOI Library (http://library.doi.gov/ill.html)
- U.S. Fish and Wildlife Service Conservation Library (http://library.fws.gov)

References

Butler, Bret W. and Alexander, Martin E., eds. 2005. Eighth international wildland firefighter safety summit: human factors—10 years later. April 26-28, 2005; Missoula, MT. Hot Springs, SD: The International Association of Wildland Fire. Available: http://www.iawfonline.org/summit [July 6, 2007].

Keller, Paul, technical writer-editor. 2004. Managing the unexpected in prescribed fire and fire use operations: a workshop on the high reliability organization. Santa Fe, New Mexico, May 10-13, 2004. Gen. Tech. Rep. RMRS-GTR-137. Fort Collins, CO: U.S. Department of Agriculture, Forest Service, Rocky Mountain Research Station. 73 p. Available: http://www.wildfirelessons.net/HRO.aspx [July 6, 2007].

Putnam, Ted. 1995. Findings from the wildland firefighters human factors workshop. Improving wildland firefighter performance under stressful, risky conditions: Toward better decisions on the fireline and more resilient organizations. 12-16 June 1995; Missoula, MT. Missoula, MT: U.S. Department of Agriculture, Forest Service, Missoula Technology and Development Center. Available: http://fsweb.mtdc.wo.fs.fed.us [July 6, 2007].

TriData. 1996a. Wildland firefighter safety awareness study: Phase 1—Identifying the organizational culture, leadership, human factors, and other issues impacting firefighter safety. Arlington, VA: TriData. 202 p. plus appendices.

TriData. 1996b. Wildland firefighter safety awareness study: Phase 2—Setting new goals for the organizational culture, leadership, human factors, and other areas impacting firefighter safety. Arlington, VA: TriData. 146 p.

TriData. 1998. Wildland firefighter safety awareness study: Phase 3—Implementing cultural changes for safety. Arlington, VA: Tridata.

Weick, Karl E. 1993. The collapse of sensemaking in organizations: the Mann Gulch disaster. Administrative Science Quarterly. 38(4): 628-652.

Weick, Karl E. 2002. Human factors in fire behavior analysis: reconstructing the Dude Fire. Fire Management Today. 62(4): 8-15.

USDA Forest Service RMRS-GTR-201. 2007

ANNOTATED READING LIST

I. Human Factors and Firefighting: An Historical Perspective

The advent of human factors as an element of research in forest firefighting operations may be one of the most critical junctures in the study of firefighters and forest fire management. Most of the following articles on human factors were written by members of the fire community. This section cites key papers that led to the recognition that human factors are prime ingredients of firefighter safety. It addresses firefighter fatalities within the United States and provides practical advice from field practitioners and research scientists dealing specifically with human factors issues of firefighter safety.

Atwood, George. 1996. Attitude of wisdom: the experience component in wildland firefighter decisions. Wildfire. 5(3).

Annotation: Atwood notes that while managers must adopt a passion for safety, the true test is on the fire line. Supervisors are in charge of making decisions that ultimately put safety first. Atwood argues that improvement will not come from the top down, or the outside in, and that firefighting experience is the most critical factor in safe and effective decision making. The more experience that individuals have fighting fires, the more capable they will be at making faster and safer decisions. Fire fighting experience and commitment to safety will help create a culture in the fire fighting organization that will lead to intuitively safe decision making strategies.

Keywords: decision making, safety, firefighter safety

Braun, Curt C. 1995. Addressing the common behavioral element in accidents and incidents. In: Putnam, Ted, project leader. Findings from the wildland firefighters human factors workshop. Improving wildland firefighter performance under stressful, risky conditions: toward better decisions on the fireline and more resilient organizations. June 12-16, 1995; Missoula, MT. Missoula, MT: USDA Forest Service, Fire and Aviation Management: 28-30.

Annotation: Burning fires are a relatively constant threat to human safety. Human behavior is the critical variable in reducing the likelihood of accidents. Braun suggests five ways to address human behavior: 1) look at the entire system of events that contribute to accidents rather than restricting the scope of investigation exclusively to the immediate causes; 2) focus on frequency and look at not only reported accidents, but also "near-miss" cases; 3) evaluate current training and management techniques; 4) determine what behaviors are being reinforced formally as well as informally; and 5) establish remediation plans. Thus, in making final evaluations of accident situations and training, it is important that managers look at the entire system of behaviors contributing to the condition rather than viewing the situation in isolation.

Keywords: risk, accidents, behavior, systems thinking

Butler, Bret W. and Alexander, Martin E., eds. 2005. Eighth international wildland firefighter safety summit: human factors—10 Years Later. April 26-28, 2005; Missoula, MT. Hot Springs, SD: The International Association of Wildland Fire. Available: http://www.iawfonline.org/summit [July 6, 2007].

Annotation: The 8th International Wildlife Safety Summit continued efforts aimed at improving wildland firefighter safety by focusing on human and social factors. The summit explored a number of issues related to wildland fire safety, and the proceedings included a workshop organized around the theme "Human Factors Ten Years Later". The summit featured 28 oral presentations, six verbal presentations, and 19 poster presentations addressing a number of topics related to wildland firefighting including human factors, culture, leadership, physiological and sociological issues, legal implications, and tools and technology.

Keywords: firefighter safety, human factors

Gleason, Paul. 1991. Lookouts, communication, escape routes, safety zones. Unpublished presentation. Available: http://www.myfirecommunity.net/documents/1991_LCES_Paul_Gleason.pdf [July 6, 2007].

Annotation: In a presentation to the USDA Forest Service's national Fire and Aviation Staff, Gleason provides a clear overview of his proposed Lookouts, Communication, Escape Routes, Safety Zones (LCES) method of training firefighters for greater safety. After defining LCES, he discusses how it should be implemented on the ground. He emphasizes the importance of lookouts, span of leadership control, safety zones on downhill and indirect fire lines, floating divisions, and wildland/urban interference. He cites ignorance, casualness, and distraction as the primary causes of firefighter accidents. Supervisors must ensure that basic fire behavior and suppression skills are learned and reviewed, that shelters are not used as a safety net to justify unwise behavior, and that firefighters on the fire line monitor one another and keep communication lines open.

Keywords: LCES, training, decision making

Keller, Paul, technical writer-editor. 2004. Managing the unexpected in prescribed fire and fire use operations: a workshop on the high reliability organization. Santa Fe, New Mexico, May 10-13, 2004. Gen. Tech. Rep. RMRS-GTR-137. Fort Collins, CO: U.S. Department of Agriculture, Forest Service, Rocky Mountain Research Station. 73 p. Available: http://www.wildfirelessons.net/HRO.aspx [July 6, 2007].

Annotation: Keller's report summarizes the 4-day *Managing the Unexpected Workshop* held in Santa Fe, New Mexico, on May 10-13, 2004. This workshop focused on how to apply high reliability organizing concepts and how to overcome personal immunities to change within the prescribed fire and wildland fire use communities. A central activity of the workshop was a "staff ride," where attendees learned about the Cerro Grande prescribed fire, which escaped and burned into Los Alamos, NM, in May 2000. The staff ride allowed participants to gain insights into the application of the high reliability organizing concepts that contribute to mindfulness and ultimately enhance the capability to discover and manage unexpected events. These concepts, drawing from Weick and Sutcliffe's book, "Managing the Unexpected," include a preoccupation with failure, reluctance to simplify, sensitivity to operations, commitment to resilience, and deference to expertise (see Weick and Sutcliffe 2007 in Section III.B.2.). Participants were immersed in these concepts in order to learn how to integrate them into their future work at both the local and national levels.

Keywords: high reliability organization, organizational learning, sensemaking, Cerro Grande

Mangan, Richard. 1999. Wildland fire fatalities in the United States: 1990-1998. Tech. Rep. 9951-2808-MTDC. Missoula, MT: U.S. Department of Agriculture, Forest Service, Missoula Technology and Development Program. 17 p. Available: http://www.fs.fed.us/t-d/pubs [July 6, 2007].

Annotation: While reports on specific fatalities and fire entrapments are prepared annually, such information is typically presented without interpretations or recommendations. This summary of fire fatalities reexamines the annual report data and offers specific recommendations for improving safety and reducing fatalities. Mangan's summary notes that burnovers make up less than 20 percent of the total causes of fatalities on fires. Other sources of fatalities include aircraft accidents, heart attacks, vehicle accidents, falling snags, and other miscellaneous causes. Managers can reduce the risk of fatalities by encouraging individual responsibility, maintaining good physical fitness, ensuring continued fire behavior training, improving fire fighting strategies and tactics, and using protective clothing and equipment.

Keywords: firefighter safety

Mangan, Richard. 2002. Injuries, illnesses, and fatalities among wildland firefighters. Fire Management Today. 62(3): 36-40.

Annotation: Mangan discusses some of the physical and environmental demands placed on firefighters that contribute to injuries, illnesses, and fatalities. He makes recommendations about clothing, hydration, diet, sleep, and physical/emotional stress in order to reduce many of the potential health risks common to the firefighting profession.

Keywords: health, injury, stress

OSHA. 1995. Occupational Safety and Health Administration's investigation of the South Canyon fire, February 8, 1995. Washington, DC: U.S. Occupational Safety and Health Administration. 9 p.

Annotation: On July 7, 1994, the Occupational Safety and Health Administration (OSHA) began a formal and independent investigation into the health and safety factors leading to the deaths of one Bureau of Land Management and 13 Forest Service firefighters during the South Canyon Fire of 1994. This investigation reviews events leading to the tragedy, makes recommendations, and cites nine unsafe conditions or practices that contributed to the tragic deaths of these workers, including such things as the identity of the Incident Commander, inadequate safety zones, escape routes, weather forecasts, fire lookouts, downhill fireline construction, and various management failures.

Keywords: firefighter safety, South Canyon, OSHA

Putnam, Ted. 1995a. Findings from the Wildland Firefighters Human Factors Workshop. Improving wildland firefighter performance under stressful, risky conditions: Toward better decisions on the fireline and more resilient organizations. June 12-16, 1995; Missoula, MT. Missoula, MT: U.S. Department of Agriculture, Forest Service, Missoula Technology and Development Center. Available: http://fsweb.mtdc.wo.fs.fed.us [July 6, 2007].

Annotation: The 1994 fire season, when 34 people died, was the catalyst for this 5-day workshop, where firefighters, safety managers, psychologists, and sociologists discussed the role of human factors in firefighter safety. Workshop participants examined firefighters, firefighter crews, fire management, fire culture, and fire communities with the goal of creating a more resilient organization for firefighter safety. The workshop began with four keynote speakers (Kurt Braun, David Hart, Gary Klein, and Karl Weick) who discussed human behavior, recognition-primed decision strategies, cultural attitudes, and insights from high reliability organizations (HROs) pertinent to improving communication, leadership, group structure, and sensemaking, which in turn can decrease stress and the chance of catastrophic errors. Participants took a field trip to Mann Gulch, where they discussed, from a psychological perspective, how and why 13 firefighters died there. The final 2 days of the workshop were spent exploring solutions and developing long-term and short-term recommendations. This publication summarizes the workshop discussion. It covers four main topics: 1) reorganization strategies for fire agencies based on HRO's; 2) fire management Incident Management Team (IMT) and fire crew reorganization using crew resource management as a model; 3) better assessment and feedback, and 4) future organizational studies, changes, and training that could improve safety. It includes suggestions related to fire organizational culture, situational awareness, mission analysis, decision making, communication, leadership and cohesion, adaptability/flexibility, assertiveness, and assessment and feedback. The author notes that the workshop was a "first step in what will be a long journey toward a better understanding of the human side of wildland firefighting".

Keywords: human factors, high reliability organizing, crew resource management, situational awareness

Putnam, Ted. 1995b. Analysis of escape efforts and personal protective equipment on the South Canyon Fire. Wildfire Magazine. 4(3): 42-47.

Annotation: This article is a detailed reconstruction of firefighter behavior and personal protective equipment use on the South Canyon Fire. Putnam reveals two failures that contributed to the deaths of 14 firefighters. First, many firefighters held onto their tools and packs during the escape effort. This slowed them down and prevented many of the firefighters from escaping. Second, fire shelters were improperly deployed, lost after deployment due to turbulent wind conditions and hot gasses preceding the flames, or not deployed at all. Putnam makes several recommendations to prevent these conditions from recurring on future fires. First, firefighters must be trained in such a way as to resist the urge to carry tools when running from a life threatening fire. They must also be trained in proper shelter storage, deployment, and use through mandatory fire shelter training courses and mandatory refresher training courses.

Keywords: South Canyon, fire shelters, tools

Putnam, Ted. 1995c. The collapse of decision making and organizational structure on Storm King Mountain. Wildfire 4(2): 40-45.

Annotation: Putnam notes that the need for a serious examination of psychological and sociological factors related to firefighter safety under stressful and risky conditions is long overdue. After looking at the typical firefighter's decision making process under normal conditions and then examining another model under stressful and risky conditions, he examines the collapse of team cohesion, leadership, and decision making in the South Canyon Fire of 1994. He states that the 10 Standard Fire Orders, 18 Watchout Situations, and nine Downhill/Indirect Line Construction Guidelines may contribute to an information overload for firefighters on the job and should be reduced. Fire managers need to be more aware of current research on group cohesion, leadership, and individual decision making prior to sending firefighters

into dangerous situations. In addition, Putnam argues that current methods of training firefighters are inadequate for the stresses of actual fire fighting because they focus too much on how to avoid entrapments and too little on how to behave when they are entrapped. Training must focus more on escape, shelter deployment, and decision making under stress with an emphasis on current psychological and social science research to avoid future entrapment fatalities.

Keywords: human factors, decision making, leadership, South Canyon, Storm King

Saveland, J.M. 2005. Integral leadership and signal detection for high reliability organizing and learning. In: Butler, Bret W.; Alexander, Martin E., eds. 2005. Eighth International Wildland Firefighter Safety Summit: Human Factors—10 Years Later. April 26-28, 2005; Missoula, MT. Hot Springs, SD: The International Association of Wildland Fire.

Annotation: Saveland's analysis shows that, from 1933 to 2003, the average number of fire entrapment fatalities has not decreased, but has stayed at about 4.6 deaths per year. In order to lower that statistic and make "significant advances" in the prevention of future fatalities, Saveland suggests using the concept of *signal detection theory* and developing leaders who possess the personal humility and professional will to face adaptive challenges. *Signal detection theory* allows the analysis of critical fire safety issues and fatalities to be viewed as *hits*, *false alarms*, *misses* or *correct rejections*. Currently, most fatality investigation reports are viewed only in a simplistic, cause and effect manner, usually screened through such templates as the Ten Standard Firefighting Orders or the Eighteen Watch-out Situations. The nuances that actually cause a specific fatality or accident are lost by using such a black and white approach. Saveland demonstrates how signal detection theory nicely fits within the theoretical frameworks of other sensemaking concepts such as the learning organization and high reliability organizing. Saveland concludes this paper with a thorough discussion of various styles of adaptive leadership, including the leadership research of Ronald Heifetz (see Heifetz 1994 in Section II.D.1.), Daniel Goleman (see Goleman 2000 in Section II.E.2.; Goleman and others 2002 in Section II.D.1.), and Ken Wilber (see Wilber 2000 in Section III.E.2). To make a "quantum leap" in lowering firefighter fatalities, leaders must create an organizational climate "where the truth is heard and the brutal facts confronted". In the last section of this paper, Saveland lists some "next steps," including using signal detection theory in After Action Reviews, benchmarking the Marine Corp Warrior Project, moving the arts into safety analysis, and developing organizational policies that facilitate forgiveness and grace.

Keywords: signal detection theory, high reliability organizing, leadership, firefighter safety

Saveland, J. 1995. Creating a passion for safety vs. management oversight & inspection. Wildfire. 4(3): 38-41.

Annotation: In this hard-hitting paper, Saveland takes issue with the Occupational Safety and Health Administration's (OSHA) report prepared after the South Canyon fire (OSHA 1995). He states that "the report will not help prevent future loss of life," and that the report is, in fact, a quick fix (single-loop learning) that will not create a climate where employees have a renewed passion for safety. Saveland uses the concepts of organizational learning, especially systems thinking, to develop his critique of the OSHA report. He notes that the highest quality behavioral changes occur in employees after an accident happens, when the process of change comes from the inside-out, where the employee is allowed freedom to respond to the information produced by an accident and adjust his/her behavior accordingly. Instead, Saveland states that the OSHA report was from outside-in, that the report blamed the firefighters at South Canyon for the fatalities, and argues that this top-down blaming approach is seriously flawed. He also demonstrates how the casual factor and the nine unsafe conditions or practices analyses that were used in the OSHA report removed individual responsibility and replaced it with a blame-based patriarchal system that would only make matters worse. Saveland's method of improving this blame-based organizational culture is to start looking outside of the box, into such areas as sensemaking (see Weick 1993 in this section), how we can become highly skilled at being incompetent (see Argyris 1990 in Section II.E.1.), to study creative decision making that is value focused, and to more fully understand the power of conversational dialogue as a valid method to comprehend these complex problems (see Isaacs 1993 in Section II.A.2.). Saveland ends his paper by noting: "You can't mandate a passion for safety, it must come from the heart". Only individual responsibility created in organizations that welcome "self-responsible actions" will cause a true passion for safety.

Keywords: South Canyon, OSHA, firefighter safety, systems thinking

Thackaberry, J. A. 2005. Wisdom in the Lessons Learned Library: work ethics and firefighter identities in the Fire Orders. In: Butler, Bret W. and Alexander, Martin E., eds. 2005. Eighth International Wildland Firefighter Safety Summit: Human Factors—10 Years Later. April 26-28, 2005; Missoula, MT. Hot Springs, SD: The International Association of Wildland Fire.

Annotation: Based on fire accident investigation reports and the Standard Fire Orders, Thackaberry used ethical theory to investigate how various "work ethics" embedded in the reports were signals that the Fire Orders were changing based on changing ethical stances in the fire community. She believes the documents housed in such depositories are not only cause and effect analyses of what went wrong on fatal forest fires, but also contain texts that carry a moral force which is often contradictory. Her study was designed to "uncover the operative ethics that have justified the Fire Orders at different points in time". She uses the Blackwater Canyon in Idaho, Ludlow fire in Mississippi, South Canyon fire in Colorado, and Cramer fire in Idaho, among others, as case studies to test her ethical theory. She discussed three types of ethical systems: the duty ethic, the ethic of virtue, and the utilitarian ethic. Under the *duty ethic*, a firefighter can be fairly certain of what is "right" and what is "wrong". However, with the *ethic of virtue*, it is nearly impossible to ascertain the logic of each situation and one must depend on the decision maker to do that. The *utilitarian ethic* highlights the consequences of an action as only being good or bad if its outcomes are seen as good or bad. In the 1930s, before the Fire Orders were issued, fire operations probably worked under the virtue ethic, where firefighters modeled their job behaviors on exemplary role models. In the late 1950s, when the Fire Orders were first developed, the virtue ethic started evolving into a duty ethic where the Fire Orders were viewed more as military commands than as generalized ways of thinking about fires. The fire reports from the 1980s and 1990s are viewed by Thackaberry as fully shifting to a duty ethic posture where the Fire Orders should never be bent. In the 1990s, the Fire Orders began drifting back to a virtue ethic. She discusses veteran firefighter John Krebs' "back to basics" reordering of the Fire Orders, suggesting the reordering is representative of the virtue ethic for firefighting organizations. Thackaberry finishes with the proposition that these historic texts, if retooled from a humanistic standpoint, might be used as exemplars for high reliability organizations.

Keywords: Standard Fire Orders, ethics, communication, South Canyon, systems thinking

TriData. 1996a. Wildland firefighter safety awareness study: Phase 1—identifying the organizational culture, leadership, human factors, and other issues impacting firefighter safety. Arlington, VA: TriData. 202 p. plus appendices.

Annotation: Following the tragic deaths of 14 firefighters in the 1994 South Canyon fire, the USDA Forest Service commissioned the TriData Corporation to investigate the effects of culture, leadership, and human factors on wilderness firefighter safety. Phase I is the first of three detailed reports delivered to the Forest Service. This study is built on information gleaned from personal interviews, focus group sessions, and survey responses of over 1,000 participating wilderness firefighters across the nation. Phase I defines the problem of firefighter safety, as expressed by firefighters themselves. It focuses on six major areas: strengths of the organization, organizational culture, leadership and accountability, human psychological factors, external influences, and prioritizing the responses. The study reinforces the idea that human factors have a large impact on firefighter safety.

Keywords: firefighter safety, human factors, leadership, culture

TriData. 1996b. Wildland firefighter safety awareness study: Phase 2—setting new goals for the organizational culture, leadership, human factors, and other areas impacting firefighter safety. Arlington, VA: TriData. 146 p.

Annotation: Following the tragic deaths of 14 firefighters in the 1994 South Canyon fire, the USDA Forest Service commissioned the TriData Corporation to investigate the effects of culture, leadership, and human factors on wilderness firefighter safety. Phase I is annotated in this bibliography. Phase II is the second of three detailed reports delivered to the Forest Service. In order to promote safety, it lays out a plan for creating the "organizational culture of the future". The first of nine chapters is an introduction to the organization and scope of the report. Chapter 2 explains the methodology and statistics used for the report, and Chapter 3 discusses the highest and lowest rated solutions that resulted from Phase I of the study. The next four chapters are organized by content area and discuss the research teams' findings regarding: organizational culture, leadership and fire management, human and psychological factors, and external influences on safety. Chapter 8 provides a summary of the report and implications for future research. Finally, Chapter 9 explains how to evaluate and measure safety levels both globally and intermediately. The report contains many tables and figures to illustrate the concepts discussed.

Keywords: firefighter safety, human factors, leadership, culture, firefighter safety

TriData. 1998. Wildland firefighter safety awareness study: Phase 3—implementing cultural changes for safety. Arlington, VA: Tridata.

Annotation: Following the deaths of 14 firefighters in the 1994 South Canyon fire, the USDA Forest Service commissioned the TriData Corporation to investigate the effects of culture, leadership, and human factors on wilderness firefighter safety. The TriData study is built on the information taken from personal interviews, focus group sessions, and survey responses of over 1,000 participating wilderness firefighters across the nation. Phase III is the last of three detailed reports delivered to the Forest Service. Phase III focuses on four major areas related to wildland fire safety: organizational culture, leadership and fire management, human and psychological factors (including training), and external influences on safety. The study also includes practical advice for moving forward and implementing changes, including on-the-job training and decision skills training.

Keywords: firefighter safety, culture, cultural change, human factors, leadership

Weick, Karl E. 1993. The collapse of sensemaking in organizations: the Mann Gulch disaster. Administrative Science Quarterly. 38(4): 628-652.

Annotation: Weick reviews the details of the Mann Gulch disaster as presented in Norman MacClean's book, "Young Men and Fire". He frames the disaster as an example of disintegration of the role structure and sensemaking in an organization. At the most crucial point in the fire, firefighters on the team leader's crew were unsure of their situation and their role in the group. This situation caused them to panic. Weick uses this example to argue that organizations are held together by communication, and it is more tenuous than we think. Therefore, organizational members need to work on resilience in order to endure tragedies. Weick explains four sources of resilience: improvisation and bricolage, virtual role systems, the attitude of wisdom, and respectful interaction.

Keywords: crisis, sensemaking, Mann Gulch

Weick, Karl E. 2002. Human factors in fire behavior analysis: reconstructing the Dude Fire. Fire Management Today. 62(4): 8-15.

Annotation: Weick reviews lessons learned from the tragic Dude Fire. Weick explains how humans make sense of their experiences by talking, and that safety is best achieved when it is easy for people to make sense of the situation. Because the environment of firefighting is often uncertain, firefighters need to frequently update their interpretations of what is happening. However, people can be overwhelmed by a series of events and make mistakes referred to as the "human factors" of firefighting. He gives a detailed explanation of three human factors, including regression to habitual behaviors, tunnel vision, and misunderstanding. Weick then deconstructs the Dude Fire and discusses all the possible barriers to effective communication during that event. He concludes with practical advice on how to avoid future firefighting tragedies.

Keywords: crisis, sensemaking, human factors, Dude fire, communication, behavior

II. Foundations for Understanding Organizations

The books and articles annotated in this section will help readers understand how organizations work. By reading these works, fire managers can gain insights into such topics as how decisions are made or what kind of leaders perform best in teams. As a whole, these readings provide a foundation for understanding the dynamics of human interaction in organizational contexts. There are six subsections in this section. First, *Decision Making and Sensemaking* examines how decisions are made in organizations and explores the sensemaking perspective as an alternative to more traditional models of decision making. The second subsection, *Organizational Culture*, demonstrates how the culture of an organization shapes both decision making and sensemaking. The third subsection, *Identification and Identity*, examines how professional identity and other allegiances influence the way people act in an organizational context. The fourth subsection, *Leadership and Change*, explores the attributes of effective leaders, especially during periods of organizational turbulence or change. The fifth subsection, *Organizational Learning*, overviews recent thought on how organizations learn from past experiences. The sixth subsection, *Team and Crew Dynamics*, looks specifically at team dynamics and describes the fundamentals of effective teams.

A. Decision Making and Sensemaking

It goes without saying that decision making is a critical skill used by all fire managers at whatever level of the organization they are operating. But what is a good decision? How does one make a "good" decision? How can one tell if their decision making process is flawed? This section explores good as well as poor decision making practices.

This section provides overviews of decision making approaches (Montgomery and others 2005; Schneider and Shanteau 2002) and decision analysis (MacGregor and González-Cabán, in press). It explores what constitutes effective decision making, provides specific approaches to decision making (Bullis 1993; Klein 2003; Klein and Weick 2000; Kofman and Senge 1993; Nutt 2002; Useem 2006), and addresses how decisions are made when decision makers are faced with pressures such as budget cutbacks (Bozeman and Pandey 2004), time constraints, high stakes

outcomes (Gonzales 2004), high-levels of personal responsibility, limited information, changing goals, shifting conditions (Klein 2000), and differing management styles (Kuhn and Poole 2000; Nutt 1999).

This section then turns to readings that offer sensemaking as an alternative to traditional notions of how decisions are made. After defining sensemaking (Weick 1995), readings address how sensemaking plays out in crisis situations (Weick 1988) and the connection between decision making and safety.

Author's Picks

- Klein, Gary. 2000. **Sources of power: how people make decisions.** Cambridge, MA: MIT Press. 338 p.
- Montgomery, Henry; Lipshitz, Raanan; Brehmer, Berndt, eds. 2004. **How professionals make decisions.** Mahwah, NJ: Lawrence Erlbaum Associates Publishers. 472 p.
- Nutt, Paul C. 1999. **Surprising but true: half the decisions in organizations fail.** Academy of Management Executive. 13(4): 75-89.
- Weick, Karl E. 2001b. **Making sense of the organization.** Malden, MA: Blackwell Publishers, Inc. 483 p.

1. How Do People Make Decisions?

Bozeman, Barry; Pandey, Sanjay K. 2004. **Public management decision making: effects of decision content.** Public Administration Review. 64: 553-564.

Annotation: This article investigates whether the content of decision tasks or the reasons for decisions influence the decision making process. Specifically, the article analyzes budget cutbacks and information technology as decision content that may affect decision making. The results suggest that the content of decisions that need to be made determines the subsequent decision making process. For information technology decisions, cost effectiveness is not as important when making decisions, average decision time is much longer, and decisions are generally viewed as permanent and stable. For cutback decisions, cost-effectiveness is very important when making decisions, decisions are made much more quickly, and decisions are viewed as unstable and changeable.

Keywords: decision making

Bullis, Connie. 1993. Organizational values and control. In: Conrad, Charles, ed. The ethical nexus. Norwood, NJ: Ablex: 75-100.

Annotation: Using the USDA Forest Service as an example, Bullis shows how decision making is largely determined by decision premises. Examples of premises include values, beliefs, and more broadly, culture. Bullis explores the ability of organizations to control decision making through employee identification with shared value premises. The results indicate that professional identities influence decision making and professional relationships. This chapter presents a case study focused on three professional groups in the Forest Service—foresters, engineers, and biologists. Foresters, whose professional identities are closer to the official Forest Service identity, report organizationally "correct" mission statements and decision premises. They also consider the Forest Service as a whole, rather than their own profession, when making decisions. Foresters' levels of identification are higher than other professionals, and they make decisions consistent with official Forest Service goals. Engineers, whose professional identities are not strongly integrated with the Forest Service identity, view mission statements literally, emphasize their professional rather than organizational values, and are more likely to disagree with Forest Service values. Their disagreements in values often center on cost-effectiveness. Biologists, whose professional identities are significantly lower than foresters and engineers, were found to contradict the Forest Service mission statement. Based on a feeling that the Forest Service favors some resources over others, they try to balance decisions made in the organization. Biologists report feeling left out of decision making processes, and they also report having their decisions reversed. Thus, different professional premises affect how well individual decisions match organizational values.

Keywords: decision making, identification, Forest Service, values, relationships

Eisenhardt, Kathleen M. 1989. Making fast strategic decisions in high-velocity environments. Academy of Management Journal. 32: 543-576.

Annotation: In order to make reliable decisions, organizations need to have reliable access to critical information. However, high velocity environments place severe restrictions on information flow. Despite this handicap, many organizations in high velocity environments are still able to achieve relatively high levels of success. This study targets the micro-computer industry, which exists in an environment of continuous change, in order to learn how fast decisions are made and how decision speed connects with

performance. Results show that fast-paced decision makers use more information to make decisions and create more alternatives than slow-paced decision makers. Results also indicate that leaders who involve experienced counselors from a variety of levels in the organization in the decision making process make decisions faster than those who invest decision making power only in top-level management. In addition, while past studies find that the existence of conflict slows the speed of decisions, this study finds that conflict resolution has more influence on decision making speed than whether or not a conflict exists. Fast decision makers confront and effectively resolve conflict throughout the decision making process. Lastly, decision makers that looked for ways to integrate decisions that impacted one another were faster than those that did not.

Keywords: decision making, high velocity environment

Endsley, Mica. 1995. Toward a theory of situation awareness in dynamic systems. Human Factors. 37(1): 32-64.

Annotation: Endsley examines the interrelation between situation awareness and individual and environmental factors. Endsley argues that situation awareness is more than a person just being aware of the situation. Rather, it is a complex understanding of the situation, and requires the ability to project what the future might bring. Further, situation awareness "presents a level of focus that goes beyond traditional information-processing approaches in attempting to explain human behavior". A model of situation awareness is provided to show how various factors affect the dynamic decision making process. The author then goes on to give more technical and expanded descriptions of the factors and their interrelations with decision making and situation awareness.

Keywords: situational awareness, decision making

Gladwell, Malcolm. 2005. Blink: the power of thinking without thinking. New York: Little Brown and Company. 288 p.

Annotation: Consistently successful decision making can make or break an individual or an organization. Perhaps counter intuitively, individuals who repeatedly make effective, successful decisions are not necessarily those who have the most information or spend the most time weighing the decision. Instead, they have perfected the art of "thin slicing," by identifying in a fraction of time the key factors in a situation that are relevant. Gladwell takes an in-depth look at how the brain makes decisions and why certain people consistently make successful decisions. Individuals who make instantaneous decisions or who desire to make more effective decisions will appreciate this book.

Keywords: decision making

Gioia, Dennis. 1992. Pinto fires and personal ethics: a script analysis of missed opportunities. Journal of Business Ethics. 11: 379-389.

Annotation: This article gives a personal account of Gioia's experience with the Ford Motor Company during a defective vehicle recall. As the recall coordinator, Gioia had to manage a busy office and keep up with a large amount of information that he needed to process. In comparing himself to a fireman, he states that his job felt like putting out the big and small "fires" that came through his office. Using his experience with the recall process with Ford's Pinto in the late-1970s, Gioia articulates the way script schemas "provide a cognitive framework for understanding information and events as well as a guide to appropriate behavior to deal with situations faced". Using the Pinto car fires as a case study, Gioia discusses how organizations use scripts and schemas to describe experiences and how employees use them to make individual choices. He ends the article by asking how moral considerations, which are usually not included in scripts, can be incorporated into them.

Keywords: ethics, communication

Gonzales, Laurence. 2004. Deep survival: who lives, who dies and why. New York: W. W. Norton & Company. 320 p.

Annotation: Gonzales attempts to answer the question of why, in life threatening events, do some people survive and others die? In a series of true-life stories about people who have had skills and behaviors of "miraculous endurance" or who have met "sudden death," Gonzales describes how people get into life threatening jams and how some of them survive and how some of them roll over and die. The book is divided into two parts, the first is titled "How Accidents Happen" and the second part, "Survival". Using not only anecdotal stories, but research from neuroscience, psychology, and decision making, Gonzales furthers understanding of the often-times deadly experiences and decision making survival techniques of pilots, sailors, and mountain climbers, among many others. Showing how the "stages of survival" are important, he describes how many survivors had to deviate from plans and to improvise, on the fly, methods of surviving. This is an excellent book to benchmark one's own risky endeavors against, see how one would survive and endure, benchmark one's mental agilities and resilience against those who have survived, and begin a personal process of creating one's own repertoire of survival techniques.

Keywords: survival, improvisation, decision making

Isaacs, William. 1999. Dialogue and the art of thinking together. New York: Currency. 448 p.

Annotation: In this book, Isaacs argues that we should learn to kindle and sustain a new conversational spirit in our relationships, organizations, and communities. A conversational spirit will help us in times of frustration and conflict and help us work together to a more promising future. Isaacs discusses what encourages and discourages such dialogue, what happens when you need to talk through difficult situations, and how to become more open to engaging in dialogue. The book is organized into five parts. The first part of the book defines dialogue. The second part concerns building personal capacity to engage in dialogue including chapters on listening, respecting, suspending judgment, and voicing opinions. Isaacs writes about intuition in the third part of the book, and in the fourth part of the book, he talks about the "architecture of the invisible," or how to make time and space for conversation to happen. Finally, the fifth part of the book concerns widening the dialogue circle and invigorating civic engagement and democracy in our communities.

Keywords: dialogue, conflict, listening skills, relationships

Janis, Irving L. 1972. Victims of groupthink: a psychological study of foreign-policy decisions and fiascoes. Atlanta, GA: Houghton Mifflin Company. 277 p.

Annotation: According to Janis, groupthink occurs when "members of any small cohesive group tend to maintain esprit de corps by unconsciously developing a number of shared illusions and related norms that interfere with critical thinking and reality testing". Groupthink can lead to group members making faulty decisions together because they value unanimity over the consideration of all possible alternatives. Using a combination of social psychology, history, and political science, Janis tests his groupthink hypotheses against four major historical events that he labels "major fiascoes". Then, showing how groups and organizations can stave off groupthink, he describes two historical case studies where decision makers constructed realistic appraisals and properly evaluated the consequences without succumbing to groupthink. The four fiascoes analyzed against groupthink theory are: President Roosevelt's failure to anticipate the attack on Pear Harbor; President Truman's decision to invade North Korea; President Kennedy's Bay of Pigs invasion in Cuba; and President Johnson's escalation of the Vietnam War. His two case studies of where successful give-and-take deliberation occurred instead of groupthink are the 1962 Cuban missile crisis and how President Truman's administration evolved the Marshall Plan after World War II. In the final two chapters, Janis describes what type of people and organizations succumb to groupthink and why. He discusses the timing of groupthink as teams of problem solvers move through the various stages of an issue. The last chapter offers tips on how to prevent groupthink. These include adding a devil's advocate to each team, getting feedback on the group's current thinking from outside the group and why the team does not want to quickly come to consensus, and giving the decision that is being proposed a chance to be reviewed quickly one last time from a different perspective. This book has value for managers and leaders who seek to understand the pros and cons of group decision making.

Keywords: groupthink, decision making, teams, group decision making, groups, politics

Kaplan, Robert S.; Norton, David P. 1992. The balanced scorecard: measures that drive performance. Harvard Business Review. January-February: 71-79.

Annotation: This article introduces the reader to Kaplan and Norton's "balanced scorecard" system, a set of measures designed to give a manager an overview of business performance. The scorecard includes four measures. The first measure is financial, the traditional measure of performance. The other three measures include customer satisfaction, internal processes, and innovation and improvement activities. The balanced scorecard prevents information overload and forces the manager to focus only on these four critical indicators of performance. Kaplan and Norton give an overview of each measure and illustrate each one with anecdotes from their research. They argue that performance measures have historically had a "control bias," and that implementing the balanced scorecard system challenges this bias by making vision and strategy central to good performance. Through the use of four measures, the balanced scorecard helps managers better understand internal relationships, which can help improve decision making and problem solving.

Keywords: decision making, strategy, performance measures, relationships

Kaplan, Robert S.; Norton, David P. 1993. Putting the balanced scorecard to work. Harvard Business Review. September-October: 1-14.

Annotation: This article summarizes Kaplan and Norton's earlier work on the "balanced scorecard" system, a set of measures designed to give a manager an overview of business performance. This comprehensive scorecard system is grounded in an organization's strategic objectives and competitive demands. It includes measures of financial success, customer satisfaction, internal processes, and innovation and improvement activities. By choosing a few indicators in each measurement area, the scorecard shows a balanced picture that can help an organization focus on its unique vision. To illustrate the use of the scorecard, Kaplan and Norton provide three case studies in different organizations including Rockwater, an engineering and construction company, Apple Computers, and Advanced Micro Devices, a semiconductor company. They conclude that the balanced scorecard is not only a measurement system, but a management system that is particularly effective in navigating change.

Keywords: decision making, leadership, performance measures, strategy, organizational change

Klein, Gary. 2000. Sources of power: how people make decisions. Cambridge, MA: MIT Press. 338 p.

Annotation: Klein presents observations of humans acting under real-life constraints such as time pressure, high stakes outcomes, high-levels of personal responsibility, limited information, changing goals, and shifting conditions. Klein studies decision making in the field—observing firefighters, intensive-care units, and chess games—to learn how people make choices when faced with constraints and difficult situations. Klein's book presents an overview of the research approach of *naturalistic decision making* and expands our knowledge of the strengths people bring to difficult tasks. Naturalistic decision making is based on the idea that all humans have the capacity to develop, through experience, the skill sets experts call upon to make good decisions. According to Klein, decision making favors both the experience of the decision maker and the context in which the decision making process is meaningful. A new context can change the way that decisions are made, so it is important to understand all characteristics that make the context unique, such as time pressure or limited information.

Keywords: naturalistic decision making

Klein, Gary. 2003a. Not all decisions are created equal: when faced with a series of tough choices, where do you start? Across the Board. March-April: 22-25.

Annotation: Problem solvers need to examine the differences that exist between decisions and the approaches available for making decisions. This short article presents four types of decisions problem solvers face and offers recommendations for each. These types of decisions include: zone of indifference choices, comparison choices, intuitive choices, and problem-solving choices. *Zone of Indifference* decisions happen when one choice is not better than the other. The traditional coin toss is usually a good way of approaching this kind of choice since it requires subsequent action. *Comparison choices* require an analytical approach because decision makers need to compare options based on common criteria. *Intuitive choices* require some level of past experience with similar situations to make the right choice. *Problem-solving choices* encourage discussion and negotiation of desired outcomes. For this, it is important to delegate responsibilities throughout the decision making process since problem-solving may take substantial time. Managers can use this article to measure how well they are approaching the choices they must negotiate on a daily basis.

Keywords: decision making

Klein, Gary. 2003b. Intuition at work. New York: Currency Books. 288 p.

Annotation: Intuition is an important factor in decision making, equal to the roles of reading data and interpreting numbers. Klein defines intuition as "the way we translate our experiences into action". Based upon his research, involving interviews with a number of life-and-death decision makers, Klein found that 90 percent of critical decisions are based upon intuition. This book discusses how intuition is not only a necessary skill, but a learnable one. Klein then offers advice on how to not only strengthen the skill of intuition, but how to apply it in the workplace. There are three main sections to this book covering first, how to build intuition, second, how to apply learned intuition, and third, how to maintain effective intuition decision making practices. Through its use of examples, practice scenarios, and exercises, this book is useful for managers, leaders, and individuals involved in a decision making process—particularly those who make decisions under extreme circumstances.

Keywords: intuition, decision making, organizations

Klein, Gary; Weick, Karl E. 2000. Decisions: making the right ones. Learning from the wrong ones. Across the Board. June: 16-22.

Annotation: This article critiques two predominant forms of decision making: rational-choice and the intuitive approach. The authors suggest a more productive approach to decision making is the experiential, or "recognize/react," approach. The experiential approach asserts that experience provides decision makers the ability to size up situations, recognize ways of reacting to situations, mentally map out options to see if they will work, focus on most relevant information, form expectancies, detect problems, and figure out ways to explain unusual events. Mental conditioning and experience help people become better decision makers and are an important part of this approach. For those who are serious about becoming proficient decision makers, the authors offer steps for mental conditioning. Steps include: defining decision requirements (time pressures, constraints, and so on), obtaining feedback about decisions made, making up scenarios (experiences) and using them, observing how the decision maker manages uncertainty, and taking advantage of people who have expertise.

Keywords: decision making, uncertainty

Kuhn, Timothy; Poole, M. Scott. 2000. Do conflict management styles affect group decision making? Evidence from a longitudinal field study. Human Communication Research. 26: 558-590.

Annotation: This article examines the relationship between group management styles and the effectiveness of group decision making. The researchers first identified conflict management styles and then analyzed group decisions and their effectiveness. The results of the study show that most groups develop management styles consistent with one another. Further, groups that develop integrative (or cooperative) conflict management styles make more effective decisions than groups that use confrontational and avoidant conflict management styles. Groups that do not develop stable conflict management styles are less effective than groups with integrative styles. While the study is based on theoretical knowledge, it is very useful for organizational members looking to further understand conflict management styles, decision making, and the connection between the two.

Keywords: decision making, conflict, groups, group decision making

MacGregor, Donald G.; González-Cabán, Armando. In Press. Decision modeling for analyzing fire action outcomes. Gen. Tech. Rep. PSW-GTR-XXX. Riverside, CA: Pacific Southwest Research Station, Forest Service, US Dept. of Agriculture.

Annotation: In this paper, the authors demonstrate a decision process that can be used to understand, review, and improve fire incident decision making and associated outcomes. The authors present theoretical models of decision making that account for dynamic, time-pressured decision making with multiple, interacting influences (for example, social, organizational, and incident-specific influences). They use two fires to demonstrate how incident decisions can be decomposed into discrete events for decision analysis and then reconstructed in order to understand the variety of influences to the incident's decisions and outcomes. The process relies on incident documentation (for example, Wildland Fire Situation Analyses) and guided interviews with key incident personnel to develop graphical representations of the sequential decision events and associated influences to the overall decision process. Based on the two example fire incident reconstructions, the authors conclude that decision processes vary among and within incidents, especially between initial/extended attack and ongoing incident management. The authors conclude with suggestions on how this process can be used to improve incident decision processes. For example, it can be used to identify management discontinuities that inhibit the use of local knowledge by incoming incident management teams. It can also be used to identify which incident decisions are pre-conceived legacy decisions. This process will be particularly useful for analyzing fires with significant consequences, such as the loss of life or unusual suppression costs.

Keywords: decision making, decision analysis

Montgomery, Henry; Lipshitz, Raanan; Brehmer, Berndt, eds. 2004. How professionals make decisions. Mahwah, NJ: Lawrence Erlbaum Associates Publishers. 472 p.

Annotation: This book was published following a conference on naturalistic decision making held in Stockholm in 2000. *Naturalistic decision making (NDM)* is a subset of decision making theory that focuses on situations where there are ill-structured problems; uncertain dynamic environments; shifting, ill-defined, or competing goals; action/feedback loops; time stress; high stakes; multiple players; and organizational goals and norms. The book updates NDM research originally compiled in the 1993 edited volume by G.A. Klein et al., "Decision Making in Action: Models and Methods". The 2004 book is organized into three sections: individual decision making, social decision making, and decision making methods. Readings on *individual decision making* include the second chapter, which describes the mental models that are constructed when decisions are made by "experienced people under complex dynamic and uncertain situations"; the third chapter, which focuses on command control and forest firefighting, and concludes that more information and resources don't always lead to better decisions; the second and sixth chapters, which discuss situation awareness as a critical element in some types of decision making; and the ninth chapter, which focuses on the role of organizational goals and norms in professional decision making. The *social decision making* part of the book investigates the social context around professional decision making, and includes chapters on team decision making and shared mental models. In this section, Chapter 14 addresses performance of fire ground commanders; Chapter 17 describes the role of culture in decision making; Chapter 18 analyzes the Challenger disaster from a cultural perspective; and Chapter 19 describes training methods for decision makers coping with ethical problems. Overall, this book reports on a variety of empirical research and theory by NDM researchers. Although NDM is a relatively new field, this book begins to illustrate that naturalistic decision making models are complex and they must account for many interactive influences on decision making.

Keywords: naturalistic decision making, Challenger, ethics, high velocity environment, culture

Nutt, Paul C. 1999. Surprising but true: half the decisions in organizations fail. Academy of Management Executive. 13(4): 75-89.

Annotation: According to Nutt, decisions that fail in organizations can be traced to managers who impose decisions, limit the search for alternatives, and use power to implement plans. Nutt finds that managers who make the need for action clear, set objectives, search for multiple alternatives, and encourage participation from others are more likely to succeed in their decision making. This decision making process first includes noticing signals that question an organization's effectiveness. Information is then gathered and an assessment is made as to whether or not action is necessary. If action is needed, organizational members must establish direction, identify options, develop a plan, evaluate plans against other alternatives, implement a plan, and assess whether new action needs to be taken. Understanding how the decision making process works can help managers make more reliable decisions and give them the tools they need to evaluate past decisions and target specific areas for future improvement.

Keywords: decision making

Nutt, Paul C. 2002a. Making strategic choices. Journal of Management Studies. 39(1): 67-96.

Annotation: This article offers four alternative prescriptions for making strategic choices in organizations: 1) analysis; 2) inspiration; 3) bargaining; and 4) judgment. Decision makers should use *analysis* when both the objectives and the means for producing results are knowable. This means commissioning a pilot test to evaluate means and identifying the best option. Alternatively, when the ends/objectives and the means for producing results and are unknown, decision makers should use *inspiration*; they should network with stakeholders to find what might work and then adapt to the needs and insights of key stakeholders. When the ends/objectives are unknown and the means for producing results are knowable, decision makers should use *bargaining*, where they create a group of stakeholders and ask the

group to find an agreeable option. Lastly, when the ends or objectives are known and the means for producing results are unknown, decision makers should use *judgment*, where they determine the option that can meet performance norms. To test these prescriptions, Nutt looked at ten cases where strategic decisions were being made. He found that four of those cases followed the prescriptions detailed above and these had a higher potential for success.

Keywords: decision making, decision analysis, judgment

Nutt, Paul C. 2002b. Why decisions fail. San Francisco: Berrett-Koehler Publishers. 332 p.

Annotation: In this book, Nutt discusses why half of all decisions that are made fail, how a decision becomes a fiasco, and how failures can be prevented. Failed decisions occur as a result of three blunders (rushing to judgment, misusing resources, and applying failure-prone tactics) and seven traps (misleading claims, barriers to action, lack of direction, limited search and no innovation, misusing evaluation and ignoring risk, ignoring ethical issues, and learning failures). Following a discussion of these blunders and traps, Nutt offers eight steps that decision-makers can use to increase the likelihood of success in their next decision making endeavor. Throughout, Nutt provides real-life examples of both failed and successful decisions including an analysis of what went wrong with specific failed decisions and why. Nutt also comments on why specific decisions were successful and how to apply their success to decision making. This book is directed toward mid- and upper-level managers of organizations as well as those who work with them.

Keywords: decision making, decision traps

Palmer, Ian; Dunford, Richard. 1996. Reframing and organizational action. Journal of Organizational Change Management. 9(6): 12-25.

Annotation: Palmer and Dunford analyze the concept of reframing and discuss four key limits to this concept. Reframing literature asserts that people generally are trapped into a singular way of thinking about a situation, and thus, they are unable to think more creatively about situations and problems they may encounter on the job. Some scholars advocate reframing for managers because it allows them to interpret situations differently than if they hadn't reframed them and to examine a variety of actions that might be possible within a given situation. However, the authors discuss four hindrances to reframing. First, addressing cognitive limits, the authors question whether reframing is something that comes naturally to people or whether it can be taught. Second, they argue that some frames are dominant within a particular organization and so there may not be the language or possibility to reframe situations outside the organization's limits. Third, the possibilities of what type of action may be needed (mental or physical) may limit where/when reframing can occur. Finally, the authors argue that reframing cannot occur on all levels of knowledge, and therefore, depending upon the situation and knowledge created during the reframing process, reframing itself may be counterproductive.

Keywords: reframing, decision making, leadership

Russo, J. Edward; Shoemaker, Paul J. H. 1989. Decision traps. New York: Doubleday. 304 p.

Annotation: As with any other skill, the ability to make effective decisions can be taught and improved upon. Russo and Shoemaker provide a guide to systematic decision making by delineating several key points in the decision making process. The first step in the process is to determine how to frame issues effectively. Once the problem/issue has been appropriately framed, the next step is information gathering and intelligence. Through systematically collecting and using information, managers can reduce traditional decision making pitfalls such as overconfidence and availability biases. Russo and Shoemaker also offer advice on how to improve information gathering techniques. Once the issue has been appropriately framed and intelligence regarding your problem has been gathered, section three covers how to make the actual decision. The authors also discuss decision making in both an individual and group context. Crucial to becoming an effective decision maker is learning from mistakes, thus, the authors also discuss reasons why people fall into poor decision making patterns and how to improve feedback.

Keywords: decision making, decision traps, group decision making

Schneider, S.L.; Shanteau, J., eds. 2002. Emerging perspectives on judgment and decision research. New York: Cambridge University Press. 736 p.

Annotation: This edited book is an excellent resource for those who wish to probe deeper into the state of the art research on emerging issues in judgment and decision making. The editors' stated purpose is to provide "fresh perspectives on decision making". The authors are particularly interested in how non-traditional topics such as memory, emotion, and context impact judgment and choice. The book is organized around five themes: 1) fortifying traditional models of decision making by looking at traditional topics in new ways; 2) elaborating on cognitive processes in decision making by exploring the interplay between decision research and cognitive psychology; 3) integrating affect (feelings) and motivation in decision making by relating how affect and motivation interact with decision making; 4) understanding social and cultural influences on decision making by recognizing the importance of social and cultural contexts for decisions; and 5) facing the challenge of real-world complexity in decision research through seeing the challenges and rewards of research outside the laboratory. The 20 articles included in this volume are written by some of the most creative decision researchers in the world and cover a wide variety of themes. Examples include "hard decisions,

bad decisions"; "memory and decision making"; "what do people really want"; the social cultural contexts of decision making in organizations; command style and team performance; the naturalistic decision making process, and how to tell if someone is an expert. The volume ends with a commentary on optimists, pessimists and realists, and how each of them might view the world of decision making and judgment.

Keywords: decision making, judgment, choice

Useem, Michael. 2006. The go point: when it's time to decide—knowing what to do and when to do it. New York: Crown Publishing Group. 275 p.

Annotation: In Useem's earlier book, "The Leadership Moment," he described leadership lessons that could be learned from various situations, including the Mann Gulch disaster. In this book, Useem has turned his attention from leadership to decision making, where he states that every decision comes down to a "go point—that decisive moment when the essential information has been gathered, the pros and cons weighed, and the time has come to get off the fence". Useem interviewed scores of people up and down the chain of command for this book (for example, CEO's, former U.S. Presidents, on-the-ground firefighters) and he uses numerous case studies, including the fall of Enron, the South Canyon wildfire entrapments, and General Robert E. Lee's decision making at Gettysburg, to develop his main points. This volume is rich with stories that are enjoyable to read, and each chapter ends with charts displaying decision making principles and techniques that can be garnered from a close reading of the stories in the book and then applied to the reader's work life.

Keywords: decision making, Mann Gulch, South Canyon

Vaughan, Diane. 1996. The Challenger launch decision: risky technology, culture, and deviance at NASA. Chicago: The University of Chicago Press. 592 p.

Annotation: In the wake of the explosion of the 1986 space shuttle Challenger, a conventional explanation for the tragedy emerged: the economic strain on NASA caused managers to withhold information about safety violations in order to maintain the launch schedule. In her book, Diane Vaughan contradicts this conventional explanation by providing a sociological explanation of the tragedy. Vaughan analyzes archival data and interviews to create a detailed historical account that answers why, in the years preceding the Challenger launch, NASA continued launching with a design that was known to be flawed. In addition, Vaughan also discusses why NASA launched the Challenger against the eve-of-launch objections of engineers. Vaughan's book describes how workplace cultures are created and sustained, and how they affect decision making. Vaughan also explains how mistakes and disasters are socially organized and systemically created by social structures. Over time, the result is a process Vaughan calls "the normalization of deviance". Overall, the book provides insight into macro-micro connections in organizational analysis, and meticulously explores the risks inherent in organizations that deal with cutting edge technology.

Keywords: decision making, risk, culture, technology

Weick, Karl E. 1996. Drop your tools: an allegory for organizational studies. Administrative Science Quarterly. 41(2): 301-314.

Annotation: One of the critical mistakes made by wildland firefighters during both the Mann Gulch and South Canyon fires was their unwillingness to drop heavy tools and packs as they attempted to outrun the flames. Weick points to 10 possible reasons for their unwillingness: listening, justification, trust, control, skill at dropping, skill with replacement activity, failure, social dynamics, consequences, and identity. Using tools as a springboard, Weick highlights four guiding principles that might be used to avoid mishaps: focus on relationships, use abstract concepts, bridge observations and abstractions, and express the values that matter. Although written to an academic audience, Weick's advice has direct application to fire managers who can learn from the similarities between the two fires in order to improve safety procedures.

Keywords: tools, identity, South Canyon, Mann Gulch, relationships

Weick, Karl E. 2001a. Tool retention and fatalities in wildland fire settings: conceptualizing the naturalistic. In: Klein, Gary; Salas, Eduardo, eds. Naturalistic Decision Making. Hillsdale, NJ: Erlbaum: 323-338.

Annotation: Comparing several well-known wildfires, Weick argues for a causal connection between firefighter tool retention and fatalities. To Weick, tools are an extension of firefighter identity and to drop one's tools is to let go of one's identity. He believes *improvisation* during high stress situations will increase safety and help firefighters maintain a clear identity, even if they have to leave behind the tools of their trade. Improvisation is the process of reshaping previously thought out plans by creatively applying past experience and knowledge during unanticipated situations. In other words, the best people to sustain a "can do" identity are those that can "make do" by applying what they know to a unique situation. Workers who use improvisation have more alternatives to choose from and have a greater chance of survival in dangerous circumstances than those who hold to rigid rules. Weick argues against checklists of rules guiding firefighter behavior in favor of more general guidelines that promote greater decision making flexibility and provide more opportunities for improvisation.

Keywords: improvisation, tools, decision making, identity, rules, survival

2. What is Sensemaking?

Weick, Karl E. 1995. Sensemaking in organizations. Thousand Oaks, CA: Sage. 235 p.

Annotation: Sensemaking is about how people make sense of situations. When faced with problems, people construct meaning. This constructive process plays a key role in the ultimate understanding that is developed. The meaning of a situation is both created and interpreted through sensemaking. Weick lists seven *distinguishing characteristics of sensemaking*. The process is: 1) Grounded in the multiple identities which comprise any individual; 2) Retrospective—meaning can only be assigned to that which has already happened. Thus, for any given event, many meanings are possible, so sensemaking is heavily reliant on memory and selective perception; 3) Enactive—people are part of their environments. Each time they act, the environment changes as a result. People play an active role in the circumstances in which we find ourselves, which then have an impact on them; 4) Social, contingent upon the conduct of others; 5) Continuous—sensemaking is an ongoing activity, part of an interruption in the constant flow of activity going on around them; 6) Focused on and by extracted cues. Cues become a point of reference against which organization members can act, explain, and understand. Cues help people define themselves and their positions relative to assumptions; and 7) Driven by plausibility rather than accuracy. Sensemaking is relative rather than absolute; it is based on reasonableness and instrumentality rather than truth and accuracy. In sum, once people develop an interpretation or explanation, then subsequent contrary information tends to confirm, rather than disconfirm the original explanation. In addition to explaining sensemaking, Weick notes that in organizations, ambiguity and uncertainty are two occasions that give rise to sensemaking. In situations of ambiguity the sensemaker is faced with multiple options; whereas, in situations of uncertainty the sensemaker lacks bases of interpretation. Sensemaking refers to finding meaning in small cues, discovering coherence among meanings, and checking with others to confirm or disconfirm hunches. Sensemaking is driven by both beliefs and actions. People see what they believe, and those beliefs become the basis for action.

Keywords: sensemaking, organizations, uncertainty

3. Sensemaking and Crisis

Columbia Accident Review Board. 2003. Columbia Accident Review Board Report. Vols. I-VI. Washington, DC: Government Printing Office. 341 p. Available: http://caib.nasa.gov [July 6, 2007].

Annotation: The Columbia Accident Review Board's (CAIB) investigation of the February 1, 2003 loss of the space shuttle Columbia lasted nearly 7 months. The loss of seven crew members and later, two debris searchers, lead to a thorough attempt to discover the truth behind the accident. The board recognized early that the accident was not likely to be a random event, and that it was more likely rooted in NASA's history and the space flight program's culture. Therefore, the investigation broadened to examine historical and organizational issues that may have been involved in the accident. This report, and its findings, conclusions, and recommendations, places as much weight on the historical and organizational causal factors as it does on the more easily understood physical causes. This report also discusses attributes of an organization that could more safely and reliably operate a risky space shuttle, but it does not attempt to provide organizational prescriptions. These attributes include: a robust and independent program, technical authority that has control over specifications and requirements, an independent safety assurance organization with control over safety oversight, and an organizational culture that represents a learning organization. Part I of the report deals with the accident itself, including the program, launch, accident analysis, and other factors involved. Part II discusses why the accident occurred, including a comparison to the Challenger accident, issues of decision making, organizational causes, and historical causes. Part III explores the look ahead, including implications, observations, and recommendations for the future. Part II is of particular interest to those concerned with high reliability organizations, especially Chapters 6 and 7, which deal with decision making and the accident's organizational causes.

Keywords: learning organization, high reliability organization, Columbia, decision making, culture, Challenger

Gephart, Robert P., Jr. 1993. The textual approach: risk and blame in disaster sensemaking. Academy of Management Journal. 36(6): 1465-1514.

Annotation: This article investigates responses to a gas pipeline explosion as a means of uncovering the methods that organizations and other participants use to make sense during disaster and to change situations. Sensemaking deals with how organizations and individuals explain or "make sense of" what goes on around them. Gephart is interested in how people use communication as a means of making sense of disasters. He finds that participants use communication to make sense of disaster situations in a number of ways, including conversations related to responsibility, risk, and safety; open avoidance of "blaming" participants and "allocating responsibility" to those who controlled events; and using common words and phrases found within the organization in order to explain the disaster. Looking at how people talk about disasters uncovers their values. For example, talk following the pipeline explosion centered around responsibility, risk, and safety, and it favored words and phrases commonly used in the organization. Understanding what people value allows managers to more effectively motivate them for change.

Keywords: sensemaking, disaster, risk, communication

Larson, Gregory. S. 2003. A "worldview" of disaster: organizational sensemaking in a wildland firefighting tragedy. American Communication Journal. 6(2). [No page numbers].

Annotation: From documents related to the 1994 South Canyon fire in Colorado, Larson examines how two worldviews presented by J.R. Taylor in his book, "Rethinking the theory of organizational communication: how to read an organization" function as sensemaking tools, both retrospectively and during crisis decision making. As Taylor explains, "In simulating some system and measuring its performance, one can either track the path of the entities processed by the system—the particles—or one can follow the behavior of the organizational units responsible for the processing—the activities". Larson explores documents related to the South Canyon Fire to provide evidence for a theoretical connection between these two worldviews, the individual and the organizational, and sensemaking. Specifically, Larson found that worldview perspectives discussed retrospectively may not correlate with those of the participants themselves. Additionally, he found that an organization's worldview impacts the sensemaking and decision making processes within that organization. Larson concludes by emphasizing the importance of incorporating multiple worldviews to understand sensemaking and decision making situations.

Keywords: worldview, sensemaking, decision making, South Canyon, disaster

Patriotta, Gerardo. 2003. Sensemaking on the shop floor: narratives of knowledge in organizations. Journal of Management Studies. 40(2): 349-375.

Annotation: In this study, Patriotta examines the ways in which an organization's workers experience the "everyday routines, interaction, and events that constitute both individual and social practices". He studies the narratives that are told within an organization, particularly during moments of disruptive occurrences. Examining these narratives helps explore the sensemaking tactics that organizational members use. Patriotta found that organizational members used *detective narratives* to investigate different occurrences on the shop floor and to give shape and structure to those events. Patriotta also found that the structure and identity of the team was based in the broader context of the overall organization and its principles. Through his research, Patriotta found that the organization was "brought to life" within the narratives rather than just being a presence in the life of the workers. This provides a different way of looking at the role of organizations and their implicit culture for workers.

Keywords: sensemaking, knowledge acquisition, organizations

Weick, Karl E. 2001b. Making sense of the organization. Malden, MA: Blackwell Publishers Inc. 483 p.

Annotation: This collection of Weick's writings addresses a central theme of organizational sensemaking, which he defines as a means by which organizational members retrospectively make sense of situations, actions, and choices. The first part of the book describes what sensemaking is and how it fits within the context of organizations, followed by several articles that highlight the components of sensemaking: ecological change, enactment, selection, retention, and remembering. The book concludes with several articles that offer application-based research. These address transformational change within organizations, electronic information processing, and globalization.

Keywords: sensemaking, organizations, enactment

Weick, Karl E. 1988. Enacted sensemaking in crisis situations. Journal of Management Studies. 25(4): 305-317.

Annotation: Crisis situations are often complicated by the very behaviors people use to manage crisis. Enactment suggests that how people think about the work they do shapes how they behave toward that work. This process allows people to see certain aspects of a situation while being blind to others. Enactment influences the course of a crisis situation because individuals are often over-committed to a pre-existing interpretation of an event or occurrence. Such over-commitment makes people focus on circumstances they feel they have the ability to deal with while ignoring others. This is dangerous since many of the factors leading to a crisis situation are unexpected and outside the scope of traditional training.

Keywords: crisis, enactment, behavior

B. Organizational Culture

Culture refers to shared meanings and understandings among groups of people that are developed and maintained through communication. Organizational culture is an important part of sensemaking and decision making practices in organizations. This section defines the concept of culture, describes how cultures are formed (Pacanowsky and Trujillo 1983), and discusses the importance of culture to issues of wildland firefighting safety. Building on that foundation, included readings take a closer look at how cultures impact organizations in terms of sensemaking (Weick 1983), organizational decision making, and how these influence organizational leaders and members (Weick 1985, 1991). Finally, the section explores specific organizational cultures that affect wildland firefighting organizations, including the USDA Forest Service culture and, more broadly, the wildland firefighting culture (Thackaberry 2004).

Author's Picks

- **Deal, Terrence E.; Kennedy, Allan A. 2000. Corporate cultures: the rites and rituals of corporate life.** Cambridge, MA: Perseus Books Group. 232 p. [Reprinted from 1982.]

- **Schein, Edgar H. 1999. The corporate culture survival guide.** San Francisco: Jossey-Bass Publishers. 224 p.

- **Thackaberry, Jennifer. 2004. "Discursive opening" and closing in organizational self study: Culture as trap and tool in wildland firefighting safety.** Management Communication Quarterly. 17(3): 319-339.

1. What is Culture?

Deal, Terrence E.; Kennedy, Allan A. 2000. Corporate cultures: the rites and rituals of corporate life. Cambridge, MA: Perseus Books Group. 232 p. [Reprinted from 1982.]

Annotation: In order to be successful in the modern world of work, mangers must attend to the workplace environment and strive to create strong workplace cultures. This includes paying attention to organizational values and associated traditions, rites, and rituals, as well as establishing organizational leaders. Additionally, effective managers must facilitate communication to establish a healthy workplace culture. Reading the culture or changing it is equally important in young as well as mature organizations. By learning to identify, read, and manage culture, organizational leaders can eventually reshape their organization's culture.

Keywords: culture, cultural change, morale, values, leadership, rituals, communication

Pacanowsky, Michael, E.; O'Donnell-Trujillo, Nick. 1983. Organizational communication as cultural performance. Communication Monographs. 50(2): 126-147.

Annotation: Many authors have addressed the building blocks of culture, such as meaningful metaphors, stories, values, and/or heroes. However, Pacanowsky and Trujillo suggest that more can be gained by looking at the *communicative performances* that give these building blocks meaning. *Communicative performances* involve social interaction between an organization's members. They discuss five communicative performances that establish and influence culture in organizations: rituals, "passion," sociality, politics, and enculturation. Rituals refer to unique ways of doing things that gain social significance within an organization. *Passion* comes when members talk about mundane duties in exciting ways to make their job experience more thrilling. Passion can be found in organizational story telling, metaphors, or specific language that highlights important aspects of the work experience. *Sociality* refers to specific codes of speech and behavior. For example, how a person talks and acts around a coworker is usually different than how a person talks to their manager. *Politics* deals with the ways that power, influence, and control are expressed and how they function within a culture. Finally, *enculturation* is the process that helps new members gain the cultural knowledge and skills needed to act as members of the organization. For example, many members learn through formal training what types of actions are appropriate and inappropriate at work as well as what values are preferred and rewarded. The authors argue that culture should be seen as what an organization *is* rather than something an organization *has*.

Keywords: culture, communication, organizations, politics, rituals, stories

Pettigrew, Andrew M. 1979. On studying organizational cultures. Administrative Science Quarterly. 24(4): 570-581.

Annotation: When studying organizations, researchers need to take into account not only the current conditions of the organization, but also past and future conditions. Studying the cultural processes of an organization over the long-term, rather than at single point in time, benefits both researchers and organizational members. Managers should take into consideration the events of the past, how those events have shaped the present, and how the future may be affected by the past and present to more productively manage organizational and team cultures.

Keywords: culture

Schein, Edgar H. 1999. The corporate culture survival guide. San Francisco: Jossey-Bass Publishers. 224 p.

Annotation: Corporate culture has a significant impact on corporate performance. First, Schein provides a basic overview of what corporate culture is, why it is important, how organizations can build culture, and how managers can assess their own corporate culture. The second half of this book takes an applied look at corporate culture, including an examination of the nuances of culture in young versus mature companies. Schein also addresses how to enact cultural change and manage situations in which cultures blend, such as during mergers and acquisitions. Managers, leaders, and executives who are attempting to enact change within their organization will find this book useful. Individuals looking to effect change in their organizations culture will benefit as well.

Keywords: culture, performance, cultural change

Schein, Edgar H. 1985. Organizational culture and leadership. San Francisco: Jossey-Bass Publishers. 418 p.

Annotation: The concepts of culture and leadership are closely tied together. Although the importance of culture is underplayed in leadership readings, this book posits the idea that possibly the "only thing of real importance that leaders do is to create and manage culture". Thus, leaders must understand culture and its impact on organizational life. This book is divided into three sections. The first section describes culture and its functions in organizational life. More specifically, this section emphasizes the importance of studying culture, how culture can be used to address the traditional workplace problems, and the implications of publishing organizational culture studies. The

second section focuses on the genesis of culture, including both its origin and how it has evolved. Central to these issues is an understanding of how individuals come to share the common assumptions, which are a central feature of culture. The final section addresses both cultural evolution and change. This section offers a framework for examining culture and posits the idea that culture serves a variety of functions in organizations. This book will be useful for managers who wish to understand organizational culture from both theoretical and practical angles.

Keywords: culture, leadership

2. How Do Cultures Impact Organizations?

Peters, Thomas J.; Waterman, Jr., Robert H. 1982. In search of excellence: lessons from America's best-run companies. New York: Harper & Row. 400 p.

Annotation: See annotation in Section II.D.1. *What is Leadership?* p. 31.

Reason, James. 1997. Managing the risks of organizational accidents. Brookfield, VT: Ashgate. 252 p.

Annotation: See annotation in Section III.A.3. *How do managers deal with risk/uncertainty?* p. 54.

Ruchlin, Hirsch S. 2004. The role of leadership in instilling a culture of safety: lessons from the literature. Journal of Healthcare Management. 49(1): 47-58.

Annotation: Ruchlin discusses macro-level concepts such as organizational culture and the influence of leadership on culture through the competing lenses of *normal accident theory* and *high reliability organization theory. Normal accident theory* posits that errors occur via system failures. Implicit to this perspective is the need for systems to create a proactive, ever-vigilant environment. Four subcultures are needed to create this type of environment: a reporting culture, a just culture, a flexible culture, and a learning culture. Alternatively, *high reliability organization theory* suggests that accidents occur as a result of incongruence between the level of complexity of the individuals who manage the systems and the systems themselves. Necessary to this perspective is the need for a safety culture that practices participative decision making, has less structured management, and emphasizes the "big picture". While this article focuses on the health care field, lessons from other disciplines are also incorporated. Ruchlin offers practical suggestions and advice for instilling a culture of safety.

Keywords: culture, leadership, safety, normal accident theory, systems thinking, high reliability theory

Thackaberry, Jennifer. 2004. "Discursive opening" and closing in organizational self study: culture as trap and tool in wildland firefighting safety. Management Communication Quarterly. 17(3): 319-339.

Annotation: Thackaberry notes that, in the wake of the South Canyon fire, the USDA Forest Service continues to maintain a strong rules-based culture of safety. This occurs despite the overwhelming call from wildland firefighters to build a safety environment where individuals are encouraged to think for themselves. This article takes a close look at the Forest Service's Wildland Firefighter Safety Awareness Study performed by the TriData Corporation and analyzes how firefighters responded to interview questions. While organizational self-study tends to provide members with an opportunity to re-think cultural norms, firefighter feedback from the study was largely reframed by the study's authors to support traditional safety norms rather than promote new ways of thinking. Thackaberry argues that the use of the 10 Standard Fire Orders, 18 Watchout Situations, and other rules-based safety standards stemming from military roots put unrealistic burdens on firefighters and serve as a scapegoat for organizational liability when a crisis occurs.

Keywords: culture, rules, firefighter safety

Weick, Karl. 1987. Organizational culture as a source of high reliability. California Management Review. 29(2): 112-127.

Annotation: High reliability organizations, such as nuclear reactors, aircraft carriers, and commercial airlines, cannot learn by trial and error because the consequences of mistakes are too severe. As a result, they must find ways of improving processes and increasing reliability that are outside of the traditional trial and error approach. Weick suggests using organizational culture to increase reliability. Organizational culture is important to establishing reliability for two reasons. First, it helps members make sense of complex situations in order to make critical decisions during moments of crisis. Second, it provides key values and beliefs that preserve overall hierarchical leadership even when organizational members must act outside of the formal chain of command. Stories can be an especially strong way of increasing and reshaping values that drive culture within high reliability organizations.

Keywords: culture, decision making, stories, high
  reliability organizations

Weick, Karl. 1985. The significance of corporate culture. In: Frost, Peter J.; Moore, Larry F.; Louis, Meryl R.; Lunberg, Craig C.; Martin, Joanne, eds. Organizational culture. Newberry Park, CA: Sage: 381-389.

Annotation: To better understand the significance of culture, Weick examines the similarities between corporate *culture* and corporate *strategy*. He explores the meaning of both terms and finds culture and strategy to be fairly interchangeable. Both grow out of past experiences and are used

to make future decisions. Additionally, both create order and meaning out of normally confusing situations by providing organizational members with important values that guide their actions. Members rarely notice culture unless they encounter situations that cannot be explained by their current beliefs.

Keywords: culture, strategy

C. Identification and Identity

Identification is the process by which people link themselves to something larger than the individual, such as a country, organization, or team. Through common values and a shared sense of meaning, individuals behave in ways that support the group with which they identify. As a result, identification can be a powerful tool for gaining compliance and developing group commitment and loyalty. This section begins with several articles that introduce how identification occurs, how identification relates to decision making, and how individual versus professional identities compare (Tompkins and Cheney 1985). The next set of readings provide a more detailed explanation of the impact of identity and identification on organizations in terms of decision making and control (Bullis 1991; Bullis and Bach 1991; Cheney and Tompkins 1987; DiSanza and Bullis 1999; Kaufman 2006; Tompkins and Bullis 1989). This section concludes by looking at how identification is created and maintained through communication (Barker and Cheney 1994; Barker and Tompkins 1994; Larson and Pepper 2003).

Author's Picks

- Ashforth, Blake E. and Mael, Fred. 1989. Social identity theory and the organization. Academy of Management Review. 14(1) 20-39.
- Bullis, Connie A.; Tompkins, Phillip K. 1989. The forest ranger revisited: a study of control practices and identification. Communication Monographs. 56: 287-306.
- Gioia, Dennis; Thomas, James. 1996. Identity, image, and issue interpretation: sensemaking during strategic change in academia. Administrative Science Quarterly. 41(3): 370-403.

1. What are Identification and Identity?

Ashforth, Blake E.; Mael, Fred. 1989. Social identity theory and the organization. Academy of Management Review. 14(1): 20-39.

Annotation: Social identity theory (SIT) suggests that people align themselves with categories (for example, organizations, religions, gender) that provide order to their social environments and give them a sense of personal identity. In this article, Ashforth and Mael discuss SIT and argue that organizational identification is a specific form of social identification, which is different from organizational commitment. According to SIT, there are a number of characteristics that contribute to organizational identification, including the distinctiveness of a group, its prestige, and an awareness of competing out-groups. There are also a number of consequences to organizational identification. First, an individual will choose to support and work for an organization that matches the individual's perceived identity. Second, organizational identification is positively associated with group development, cohesion, and cooperation. Finally, the individual will come to view their group values as the most important and distinctive. Ashforth and Mael then discuss the theoretical implications of SIT with respect to socialization, role conflict, and intergroup relations.

Keywords: identity, social identity theory, organizational identification, organizational commitment, groups

Cheney, George. 1983a. The rhetoric of identification and the study of organizational communication. Quarterly Journal of Speech. 69: 143-158.

Annotation: Many organizations actively work to get their members to identify with the organization in order to accomplish organizational goals. This article looks closely at internal documents from a variety of organizations and identifies four main methods of building identification: common ground, antithesis (uniting against a common enemy), the assumed "we," and unifying symbols. The first method, common ground, is discussed in terms of six tactics: 1) expression of concern for the individual; 2) recognition of individual contributions; 3) espousal of shared values; 4) advocacy of benefits and activities; 5) praise by outsiders; and 6) "testimonials" by employees. Cheney finds that unifying symbols play an important role in creating, maintaining, and building identification in organizations.

Keywords: identification, organizational identification

Cheney, George. 1983b. On the various and changing meanings of organizational membership: a field study of organizational identification. Communication Monographs. 50: 342-362.

Annotation: In this article, Cheney highlights the principles of organizational identification as well as its opposing process, alienation. He investigates how identifying with an organization can influence an organizational member's decisions at both conscious and semi-conscious levels. Cheney also investigates how organizational identification can be a strong source of employee motivation. This article is useful for practitioners who are interested in understanding organizational identification and its impact on decision making patterns.

Keywords: organizational identification, decision making

Cheney, George; Tompkins, Phillip K. 1987. Coming to terms with organizational identification and commitment. Central States Speech Journal. 38(1): 1-15.

Annotation: Cheney and Tompkins explore the relationship between organizational identification and commitment. They conclude that the process of identification yields behavioral commitments. However, the relationship between the two is complex, and not absolute. For example, while strong identification can lead to commitment to act, high levels of organizational identification are sometimes met with low levels of commitment. At other times, high levels of commitment still express themselves when individuals have low levels of organizational identification, for instance, when an employee does not like their work, but works hard in order to earn a steady income.

Keywords: organizational identification, organizational commitment

Scott, Susanne G.; Lane, Vicki R. 2000. A stakeholder approach to organizational identity. Academy of Management Review. 25(1): 43-62.

Annotation: Scott and Lane discuss manager-stakeholder relationship factors that influence how organizational members make sense of themselves and how they identify with the organization. Stakeholders are people who are linked to the organization, such as employees or citizens, and expect to benefit in some way from the organization's successes. This article looks at how stakeholder identity emerges in organizations. It then discusses how a manager-stakeholder relationship contributes to the ways managers can promote organizational identity. Understanding how identity emerges and how the manager-stakeholder relationship can impact those identities can help managers direct and evaluate group cohesion as well as individual loyalty.

Keywords: identity, stakeholders, identification

2. How Does Identification/ Identity Impact Organizations?

Barker, James R.; Tompkins, Phillip K. 1994. Identification in the self-managing organization: characteristics of target and tenure. Human Communication Research. 21(2): 223-240.

Annotation: Employees not only identify with organizational goals, but also with team goals and values. It is important to understand how team members manage competing loyalties between the team and the larger organization in order to understand how they behave in actual practice. Barker and Tompkins find that team members consistently identify more strongly with their fellow team members than they do with the larger organization. Managers need to keep this in mind when organizational values and decisions compete with team values and decisions. Members are more likely to follow what their team decides to do than what the organization does. However, the study also found that newer members identified with the team and the organization less than members with more seniority. This suggests that teams with mixed membership may be less likely to group together on decisions, since some feel a strong loyalty to the values and decisions of the team and organization while others do not.

Keywords: identification, teams, loyalty

Bullis, Connie. 1991. Communication practices as unobtrusive control: an observational study. Communication Studies. 42(3): 254-271.

Annotation: In this study, Bullis observes three internal Forest Service meetings to see how managers and employees use face-to-face communication to establish control. Bullis notices three forms of communication in the meetings: information dissemination, group discussions, and decision making. Managers and supervisors use face-to-face communication to emphasize, bolster, and reconstitute important values that guide future employee behavior. When formal organizational policies contradict previous organizational values with which members have identified, supervisors tend to ignore the policy and support the values. Bullis argues that supervisors and employees that identify with an organization based on shared values will have greater loyalty to the values than to the organization policy. She highlights an important limitation of value-based control practices for managers. Once an employee is committed to certain values, it is difficult to let go of them even if the organization decides to change.

Keywords: identification, values, control

Bullis, Connie; Bach, Betsy W. 1991. An explication and test of communication network content and multiplexity as predictors of organizational identification. Western Journal of Speech Communication. 55: 180-197.

Annotation: Identification contributes to the development of social networks. This study targeted three groups of incoming graduate students and distributed a questionnaire at three different times during their first year of school. They found that the more things a student had in common with another student, the more they would identify with them. For example, a person that shared interests in fishing, hunting, going to movies, and hiking with another person identified more with that person than someone that shared only one of those interests. This suggests managers who want to increase team identification should look for ways to increase the interests they share with those they lead.

Keywords: identification, teams, networking

Bullis, Connie A.; Tompkins, Phillip K. 1989. The forest ranger revisited: a study of control practices and identification. Communication Monographs. 56: 287-306.

Annotation: This study is a follow up to Herbert Kaufman's (see annotation of 2006 reprinted version, in this section)

research on Forest Ranger commitment. The authors explain that the USDA Forest Service has adopted a more rules-based form of controlling employee behavior, and employee identification with the Forest Service has decreased. This study shows that five practices have decreased employee identification with the Forest Service: 1) hiring practices have changed in favor of diversity, which has increased the variety of individual values and beliefs; 2) most of today's training comes from outside specialists whose values may compete for employee attention and influence the thinking of new employees early; 3) people are relocated far less frequently, giving the community and other organizations outside of work a stronger chance to compete for employee loyalty; 4) many employees skip "on the ground" training because of their educational qualifications, so they are less familiar with traditional Forest Service values; and 5) the widespread use of organizational symbols, such as uniforms and log cabins, has decreased. This study helps managers understand employee commitment by showing what elements have contributed to lowering commitment within the Forest Service. It also shows how the Forest Service has changed its methods of control from predominantly value-based to a more rules-based structure.

Keywords: control, identification, Forest Service, rules, training

DiSanza, James R.; Bullis, Connie. 1999. "Everybody identifies with Smokey the Bear": employee responses to newsletter identification inducements at the U.S. Forest Service. Management Communication Quarterly. 12(3): 347-399.

Annotation: Newsletters are a powerful channel for organizations to connect with employees and establish identification and loyalty. This study looks at an internal USDA Forest Service newsletter and discusses the tactics it uses to encourage employee identification with organizational values. Employees responded to the newsletter by either identifying with the values or by alienating themselves from the values of the organization. There were also many whom neither identified with, nor felt alienated from, the organization because of the newsletter and read it simply for pleasure or to glean specific information. The analysis argues that employee identification that is shaped early leads to more positive experiences and attitudes toward the organization.

Keywords: identification, Forest Service

Hogg, Michael A.; Abrams, Dominic; Otten, Sabine; Hinkle, Steve. 2004. The social identity perspective: intergroup relations, self-conception, and small groups. Small Group Research. 35(3): 246-276.

Annotation: Social identity theory offers an explanation of organizational loyalty. Social identity hinges on the concept of the "prototype" or embodiment of ideal characteristics that define one group from another. Prototypes allow individuals to categorize themselves and those around them into specific groups (for example, sex, race, nationality, occupation) in order to make sense of the world. However, prototypes also serve as a means of defining and influencing behavior. For example, as individuals categorize themselves, they simultaneously accept the behaviors, values, and beliefs that belong to the unique prototype associated with that category. Categorizing oneself in line with accepted prototypes allows individuals to reduce uncertainty, making actions and their consequences more predictable. This explanation is complicated by the reality that there are multiple ways to categorize individuals and therefore there are multiple prototypes from which to select.

Keywords: social identity theory, groups, decision making, values, uncertainty

Gioia, Dennis; Thomas, James. 1996. Identity, image, and issue interpretation: sensemaking during strategic change in academia. Administrative Science Quarterly. 41(3): 370-403.

Annotation: This study focuses on identity and image concepts that top management groups have about the type of organization in which they are involved. Identity has been described in the organizational behavior literature as "those features of the organization that members perceive as ostensibly central, enduring, and distinctive in character;" whereas, image has generally been defined as "how members believe others view their organization". In this article, Gioia and Thomas argue that it is necessary to "reconsider the assumed durability and distinctiveness of identity and image". They found that top management groups used the language of outsiders, image language, to help change the identification in an organization. They argue that in order to make substantive change in an organization, identity changes must occur, and those can be made via image changes. The authors also argue that an organization that includes flexibility within its identity will learn and adapt more quickly.

Keywords: identity, sensemaking, change, organizational behavior

Kaufman, Herbert. 2006. The forest ranger: a study in administrative behavior. Special Reprint Edition. Washington, DC: Resources for the Future Press. 249 p. [Reprinted from 1960].

Annotation: The connection between culture, decision making, and behavior is clear in this classic study of the USDA Forest Service. Based on lengthy observations and interviews with forest rangers in the field, this investigation explains the forest rangers' uncanny devotion, loyalty, and obedience to Forest Service ideals. Kaufman found that rangers closely identified with Forest Service values, which were reinforced by individual education, initial organizational training, community involvement, and through the use of standardized symbols such as for uniforms and

building architecture. He argues that identifying with core values motivated rangers to faithfully comply with organizationally approved behavior.

Keywords: identification, culture, loyalty, Forest Service, training

Kuhn, Timothy; Nelson, Natalie. 2002. Reengineering identity: a case study of multiplicity and duality in organizational identification. Management Communication Quarterly. 16(1): 5-38.

Annotation: Kuhn and Nelson provide a case study of a municipal development department and how its organizational members manage their work and their identities during conflict. Member identities are influenced by their perceived places in the communication network of the organization and by the context of situations. The ways members talk about who they are form, maintain, and change their identities, and these also form, maintain, and change the larger organizational identity. Consequently, the organization can influence ways in which teams see themselves and how they view issues of safety. As a result, team members can have both large and small effects on the ways in which an organization views itself.

Keywords: organizational identification

Larson, Gregory S.; Pepper, Gerald L. 2003. Strategies for managing multiple organizational identifications: a case of competing identities. Management Communication Quarterly. 16(4): 528-557.

Annotation: In this study, Larson and Pepper examine the communication strategies and tactics used by employees when multiple sources of identification appeal to them. They argue that communication strategies and tactics reinforce and intensify the way employees identify with sets of values. Through interviews and observations, the authors identify three primary strategies employees use to manage competing appeals for identification: *comparison*, *logic*, and *support*. First, employees often use *comparison* to verbally weigh their options between group values and they used the discussion of pros and cons to justify their choice of organizational or group identity. Second, employees *logically* justify their support and loyalty to a set of values by referring to generally accepted values as evidence. Third, selected employees base their loyalty to chosen values on the *support*, or perceived support, of others. The authors argue that verbal communication is a strong way to reinforce and change loyalties within an organization.

Keywords: organizational identification

Papa, Michael J.; Auwal, Mohammad A.; Singhal, Arvind. 1997. Organizing for social change within concertive control systems: member identification, empowerment, and the masking of discipline. Communication Monographs. 64: 219-249.

Annotation: This article looks at the reasons behind the astounding success rate of the Grameen Bank in Bangladesh.

The bank provides work loans without collateral to the poorest citizens in the country while maintaining a loan recovery rate of about 99 percent. Upper management uses formal training to instill the organizational value of aiding the poorest of the poor and gives field workers the freedom to develop their own ways of achieving that goal. After internalizing the goal, workers discipline one another to comply through peer pressure. Strong identification with organizational values, beliefs, and goals, combined with a desire to gain the approval of fellow workers, "masks" how the organization relies on workers to meet its own goals. This article examines both the benefits and pitfalls of strong identification.

Keywords: identification, organizations, training

Tompkins, Phillip, K.; Cheney, George. 1985. Communication and unobtrusive control in contemporary organizations. In: McPhee, Robert, D.; Tompkins, Phillip, K., eds. Organizational communication: traditional themes and new directions. Beverly Hills, CA: Sage Publications: 179-293.

Annotation: Tompkins and Cheney argue that concertive control is an unobtrusive form of control that allows team members to govern themselves by adhering to core organizational values. As individuals identify with an organization, they begin to make choices reflecting the interest of that organization. They are not forced to support the organization's values, instead, they support the organization's values because they want to and because they respect and value that organization as an important part of their own identity. This article also points to how organizational identification limits the alternatives an individual sees when making a decision. The more a person identifies with the values of an organization, the more the person wants what is best for the organization. This prevents them from considering alternatives that do not promote the organization's ascribed values. This can be dangerous when circumstances demand choices that fall outside traditional organizational values.

Keywords: identification, decision making

D. Leadership and Change

Leadership influences all organizations. This section defines leadership, explores the history of leadership studies, examines the factors that go into leadership, and addresses the idea of effective leadership. It also addresses how leaders influence organizational change. Change can be challenging. As a result, the literature on leadership and transition is extensive. This literature addresses how cultural change (Deetz and others 2000) and change resulting from environmental and organizational needs affect organizations (Kotter 1996). The articles included here illustrate what happens within organizations during transitional periods and suggest effective ways of coping with change.

Author's Picks

- Deetz, Stanley A.; Tracy, Sarah J.; Simpson, Jennifer L. 2000. Leading organizations through transition. Thousand Oaks, CA: Sage. 232 p.
- Heifetz, Ronald A. 1994. Leadership without easy answers. London: Belknap Press. 366 p.
- Herzberg, Frederick. 2003. One more time: how do you motivate employees? Harvard Business Review. 81(1): 86-96. [Reprinted from 1968.]
- Kotter, John P. 1996. Leading change. Boston, MA: Harvard Business School Press. 187 p.

1. What is Leadership?

Barker, Richard A. 2002. On the nature of leadership. Lanham, MD: University Press of America. 162 p.

Annotation: Often, leadership has been oversimplified in the literature. Barker rejects this trend by offering an in-depth look at leadership and the social processes contributing to leadership. Operating under the assumption that one cannot "research, examine, develop, and practice" leadership until the person understands what it is, Barker explains the fundamental nature of leadership. He begins with his analysis of where current scholarship and leadership practitioners have gone wrong. Secondly, Barker discusses leadership theories and the role of morals and ethics in leadership. He then discusses leadership in the context of organizations and common conventions about applied leadership. Barker concludes with a discussion of leadership as both a process and as a relationship. His analysis furthers the understanding of the nature of leadership.

Keywords: leadership, ethics

Burns, James M. 1978. Leadership. New York: Harper & Row. 530 p.

Annotation: Despite the vast amount of scholarship on leadership, no central unifying concept of leadership exists. By identifying patterns in both the origin and socialization of individuals that accounts for leadership, this book's purpose is to create a unifying concept. Burns identifies two basic types of leadership: the *transactional* and the *transformational*. The majority of leaders and followers today follow the *transactional* model where leaders and followers interact with one another on an exchange basis. In contrast, *transformational* leadership is more complex. The *transforming* leader takes a deeper look at the followers' needs and motives. This more sophisticated approach engages the follower. It has the potential to change "followers into leaders and may convert leaders into moral agents". Thus,

Barker also discusses a third type of leadership, *moral leadership*. This third form of leadership is rooted in "the fundamental wants and needs, aspirations, and values of the followers". This book is divided into five sections. Section one discusses both the power of leadership and the purpose and structure of moral leadership. Section two delves into the origins of leadership, examining the psychological, social, and political forces that influence leadership. Section three examines *transforming* leadership by looking at real-world examples. Section four focuses on *transactional* leadership in a variety of contexts including group, party, and legislative leadership. Section five covers both the theoretical and practical implications of leadership. In addition to providing an in-depth look at leadership, this book provides instruction on leadership in action.

Keywords: leadership, politics

Goleman, Daniel; Boyatzis, Richard; McKee, Annie. 2002. Primal leadership: realizing the power of emotional intelligence. Boston: Harvard Business School Press. 352 p.

Annotation: The authors of this book explain how leadership is primarily emotional. For this reason, it is crucial for good leaders to have emotional intelligence, which the authors call "primal leadership". Using a vast collection of data on emotional intelligence in the workplace, they argue that primal leadership drives performance within an organization. The book is written in three parts. The first part examines the power of emotional intelligence by defining such concepts as primal leadership, resonant leadership, the neuroanatomy of leadership, the leadership repertoire, and dissonant styles of leadership. The second part of the book focuses on making leaders and offers anecdotes and practical advice on motivation, developing leadership skills, and sustaining personal changes. In this part, the authors also introduce "the five discoveries" as tools for developing primal leadership capability. The final part of the book takes a broad look at how to build emotionally intelligent organizations. It includes sections on teams, visions, and sustainable change. Although the book is geared toward a management audience, the authors stress that leadership resides in all of us.

Keywords: leadership, emotional intelligence, management

Heifetz, Ronald A. 1994. Leadership without easy answers. London: Belknap Press. 366 p.

Annotation: In this book, Heifetz offers a practical philosophy of leadership. He argues that the culturally dominant view of leadership suggests the exercise of influence over others. However, a better image of leadership involves mobilizing people to tackle tough problems and articulating a clear mission. The book has four parts that all include practical examples, beginning with a part called "setting the frame" that discusses the core values of leadership and the roots of authority. The second part of the book focuses on leading with authority in both technical and adaptive situations. It investigates resources, dimensions, and applications of authority. In addition, Heifetz offers some principles of leadership including: identifying the adaptive challenge, regulating distress, directing disciplined attention to the issues, giving the work back to people, and protecting the voices of leadership in the community. The third part of the book tackles the challenge of leading without authority and in the face of deviance. Finally, the fourth part of the book talks about how to stay alive, including advice such as pacing yourself, listening to yourself, having a sanctuary, and preserving a sense of purpose.

Keywords: leadership, authority, stress, conflict

Heifetz, Ronald A.; Laurie, Donald L. 1997. The work of leadership. Harvard Business Review. January: 124-134.

Annotation: In this article, Heifetz and Laurie argue that organizations need to effectively adapt their behaviors in order to thrive in new environments. In this process, an effective leader will motivate employees to do adaptive work. However, most leaders are accustomed to problem solving, and attempt to shelter employees from the distress and conflict inherent in change. Heifetz and Laurie believe that a better response is to motivate employees by letting them feel the reality of the situation. Based on extensive research, the authors provide six principles for adaptive leadership. These include "getting on the balcony," identifying the adaptive challenge, regulating distress, maintaining disciplined attention, giving the work back to people, and protecting voices of leadership from below. Heifetz and Laurie outline each principle and then apply it to a case study in a professional services firm. The article refutes the traditional problem-solution leadership style and is a practical guide to adaptive leadership during organizational change.

Keywords: leadership, organizational change, organizational behavior, motivation

Kotter, John P. 1990. A force for change: how leadership differs from management. New York: The Free Press. 192 p.

Annotation: Organizational leadership is an important, yet little-understood topic. Through a two-part study involving a questionnaire and an analysis of "highly effective leadership" cases, Kotter discusses the differences between effective leadership and management. By comparing leadership to management, Kotter clarifies the functions, processes, structure, and origins of leadership. This book addresses both the nature of leadership and its relationship to management. Managers will find this book useful for learning the central functions of effective leadership, such as direction setting and alignment, as well as key characteristics of effective leaders, such as the ability to motivate others and to set organizational norms and values.

Keywords: leadership, management

Kouzes, James M.; Posner, Barry Z. 2002. The leadership challenge. San Francisco: Jossey-Bass Inc. 496 p.

Annotation: Kouzes and Posner recognize that effective leadership both motivates individuals and creates a climate of success. Leadership influences individual careers, organizations, and communities. The central theme in this book is that "leadership is a relationship". The book is divided into six sections. The first section covers five practices of effective leaders and the importance of credibility for a leader. The second section discusses the importance of personal values to leadership as well as the importance of modeling values as a leader. The third section covers two topics, being a forward-looking leader and engaging others in the leader's future vision. The fourth section discusses ways to create change and innovation in an organization as well as the risks associated with change. The fifth section focuses on enabling and strengthening others and how to build both goals and trust. The final section covers ways to encourage employees through recognizing others' contributions and celebrating their values and victories. This book is useful for individuals who wish to further their leadership skills.

Keywords: leadership, relationships, values

Likert, Rensis. 1967. The human organization: its management and value. New York: McGraw-Hill. 258 p.

Annotation: Likert advocates a scientific approach to studying management, emphasizing the importance of rigorous quantitative research. The book begins by exploring the organizational and performance characteristics of four different management systems. These characteristics include leadership processes, motivational forces, communication processes, interaction-influence processes, decision making processes, goal setting or ordering, and control processes. The author then argues that many managers use a system of management that is less productive than other available systems. In order to improve general management and fiscal responsibility, Likert provides practical science-based tools for managers of each type of system.

Keywords: leadership, management, motivation, communication, decision making, control

McGregor, Douglas. 1960. The human side of enterprise. New York: McGraw-Hill. 246 p.

Annotation: In this classic book, McGregor explores the common assumptions about the most effective way to manage employees. The book is written in three parts. The first part explores traditional management theory. This includes a discussion of influence and control. McGregor writes that the traditional view of control, or Theory X, rests on three assumptions: humans detest work, seek to avoid it, and need to be coerced into it. However, according to Theory Y, humans can be self-motivated and they see work as integrated, natural, and creative. In the second part of the book, McGregor explains how to manage employees using Theory Y, including tips on how to handle performance appraisals, salaries, and promotions. He stresses employee participation in the workplace, fostering collaborative relationships, and encouraging a positive climate. Finally, in the third part of the book, McGregor offers practical advice on how to train and develop managers and managerial teams.

Keywords: leadership, management, control, training, Theory X, Theory Y, relationships

Mintzberg, Henry. 1990. The manager's job: folklore and fact. Harvard Business Review. 68(2): 163-177. [Reprinted from 1975].

Annotation: This classic article challenges the idea that managers do things such as plan, organize, coordinate, and control. Instead, Mintzberg argues that managers have roles. Managerial roles are categorized as *interpersonal*, *informational*, and *decisional*. *Interpersonal* roles include acting as a figurehead, leader, and liaison. *Informational* roles include acting as a monitor, disseminator, and spokesperson. *Decisional* roles include acting as an entrepreneur, disturbance handler, and resource allocator. In order to be effective, Mintzberg also argues that managers must have insight into their own work and they must learn to share information, give attention to issues that need it, and gain control of their own time.

Keywords: management, managerial roles, decision making, leadership

National Wildfire Coordinating Group. 2007. Leading in the wildland fire service. Boise, ID: National Wildfire Coordinating Group. 68 p. Available: http://www.fireleadership.gov [July 6, 2007] or http://www.nwcg.gov/pms/pubs/pubs.htm (Order# NFES 2889; PMS-494-2) [July 6, 2007].

Annotation: This book's stated purpose is to provide a fundamental leadership framework for fire managers, fire leadership instructors, and students of fire leadership at all functional levels in the wildland fire service. It defines leadership and then describes a universal set of values and principles to guide leaders in fire operations. Noting that the fire environment is often one of high risk, where decisions have to be made swiftly under chaotic conditions, leadership is broadly defined as "the act of influencing people in order to achieve a result". The overarching leadership framework is to become effective at leading people. To accomplish this, a leader must be aware of the leadership environment, which includes the leaders themselves, the people being led, the unique variables that influence a leader's decision processes, and the short- and long-term consequences of the leader's decisions. A command philosophy must be developed where the leader's intent is clear and leads to a unity of effort. Having a positive command climate—trust, open communication, respect, and the freedom to debate issues—is also important. Four levels of leadership are discussed: followers, leaders of people, leaders of leaders, and leaders of organizations. After establishing a leadership framework, the book describes leadership in terms of three core values—*duty*, *respect*, and *integrity*. The major *duties* that a leader is responsible for are: being proficient at the work being done, making good decisions, ensuring work objectives are understood, and developing the people who complete the work. *Respect* is concerned with knowing your subordinates and looking out for their well being, keeping the crew informed, building the team, and utilizing people in accordance with their capabilities. Finally, *integrity* covers: knowing ourselves as leaders and improving our capabilities, seeking responsibility for our actions, and setting the example. The book contains numerous sidebars where examples and models are described. Sidebar examples include unity of effort, mentoring, after action reviews, moral courage, and leading up. Sidebar models describe situation awareness, decision making cycles, and responsibilities of communication. At the end of the book, a section titled "Eyes Forward," the authors discuss the lifelong quest to master the duties and principles of leadership.

Keywords: leadership, values

Northouse, Peter G. 2003. Leadership: theory and practice. Thousand Oaks, CA: Sage. 360 p.

Annotation: Organizations demand effective leadership. While numerous texts have been written regarding leadership theory, this book provides a necessary link between leadership theory and real-life application. Multiple approaches to leadership are discussed including trait, skills, style and situational approaches. Additionally, leadership theories, such as contingency and leader-member exchange theory are discussed. Both theory and application are included in the discussion of leadership approaches, as well as a debate of the merits and weaknesses of each approach. Finally, case studies are provided to illustrate leadership approach applicability. This book is useful for understanding leadership theory and its application in real world contexts.

Keywords: leadership

Parks, Sharon Daloz. 2005. Leadership can be taught: a bold approach for a complex world. Boston, MA: Harvard University Press. 287 p.

Annotation: Parks spent a year in Ronald Heifetz's leadership classroom at Harvard University's John F. Kennedy School of Government to describe his "case-in-point" teaching method. Heifetz, author of "Leadership Without Easy Answers," is acknowledged by many as one of the premier professors of leadership in the United States. Heifetz not only teaches his vision of leadership dilemmas—authority versus leadership; technical problems versus adaptive challenges; power versus progress; and personality versus presence—but actually mimics these leadership attributes, thought processes, and behaviors in the classroom through role playing, simulation, dialogue, and lecture seminars between himself and his students. The classroom becomes a real life example of what the students will face in the outside world of work. The students are taught to write critical self reflections, use emotional resilience, practice failing publicly, and understand the challenges associated with intent

listening. The case-in-point teaching method is described as rigorous, and many students drop out of the class because of the psychological pressures generated by the case-in-point pedagogy.

Keywords: leadership, dialogue

Peters, Thomas J.; Waterman, Jr., Robert H. 1982. In search of excellence: lessons from America's best-run companies. New York: Harper & Row. 360 p.

Annotation: In the wake of the Japanese manufacturing successes of the early 1980s, this book provides one of the founding arguments for culture as a means of control and achieving excellence in organizational management. After interviewing executives in the top financially successful corporations in America, such as 3M Company, Delta, IBM, Proctor and Gamble, and Hewlett Packard, the authors present eight values of management excellence commonly practiced among them: 1) a bias for action; 2) close to the customer; 3) autonomy and entrepreneurship; 4) productivity through people; 5) hands-on, value driven; 6) stick to the knitting or stay close to the businesses you know; 7) simple form, lean staff; and 8) simultaneous loose-tight properties. This book provides a wide variety of examples and models to support and explain these eight values of successful corporations. This book is useful for understanding how culture can be used to motivate and lead employees.

Keywords: culture, control, values, leadership

Saveland, J.M. 2005. Integral leadership and signal detection for high reliability organizing and learning. In: Butler, Bret W.; Alexander, Martin E., eds. 2005. Eighth International Wildland Firefighter Safety Summit: Human Factors—10 Years Later. April 26-28, 2005; Missoula, MT. Hot Springs, SD: The International Association of Wildland Fire.

Annotation: See annotation in Section I. *Human Factors and Firefighting.* p. 9.

Sloan, Diane K.; Krone, Kathleen J. 2000. Women managers and gendered values. Women's Studies in Communication. 23: 111-131.

Annotation: In this study, Sloan and Krone examined 30 women managers' responses to masculine and feminine values and power orientations. They argue that, because gendered practices exist in organizations, women managers should consider a form of networking that decreases feelings of isolation within organizations. Two suggestions from the article include learning from women who have been successful in a masculine world and a more action-oriented approach where women share experiences and offer guidance and mentoring for each other. The authors offer a starting point for new networking strategies, but they acknowledge that it is not a simple process.

Keywords: gender, leadership, networking

Tannen, Deborah. 1994. Talking from 9 to 5: men and women at work. New York: HarperCollins. 361 p.

Annotation: In this book, Tannen looks at the role of gender in communication in the workplace. Tannen covers such topics as conversational rituals and potential barriers in communication between genders in the business world. The author discusses tools to help build a more positive and productive communication climate in the workplace. Tannen focuses on the role of women in the workplace, specifically in leadership positions. To this end, Tannen covers concepts such as the glass ceiling, women and authority, and status and connection. Tannen concludes with comments on speaking in meetings at work.

Keywords: gender, rituals, leadership, authority, communication

Wren, Thomas J., ed. 1995. The leader's companion: insights on leadership through the ages. New York: The Free Press. 376 p.

Annotation: Given its increased status in popular literature, the concept of leadership is a frequently discussed, yet a little understood phenomenon. This book provides a broad, process-based approach to the phenomenon of leadership. In his discussion of leadership, Wren combines observations from different authors of scholarly articles on leadership and leader insights. This book is organized into 13 sections that range in topic from basic information about leadership to the comments of both classical and modern scholars concerning leadership. Additionally, the book discusses the process of leadership itself, including how issues such as gender and diversity impact it. This book will be useful for developing a comprehensive understanding of "the art and science of leadership".

Keywords: leadership, gender

2. How Do Leaders Create/Manage Change?

Block, Peter. 1993. Stewardship: choosing service over self-interest. San Francisco: Berrett-Koehler. 264 p.

Annotation: The central theme of this book focuses on the concept of stewardship. Block defines stewardship as "accountability without control or compliance". According to Block, stewardship effects change in an organization by operating "in service rather than in control" of organizational members. Thus, stewardship is described as a governing strategy. This book is divided into three parts. Section one discusses the concept and promise of stewardship as well as the limitations of leadership. Section two discusses the practical implications of stewardship in action. Section three describes the logical side of stewardship in action and discusses how to implement the concept of stewardship. This book will be useful for managers who want to implement a service-oriented managerial policy or who seek to enact change in traditionally difficult domains in the workplace such as the distribution of power, purpose, and rewards.

Keywords: leadership, strategy

Bolman, Lee; Deal, Terrence. 1991. Reframing organizations: artistry, choice, and leadership. San Francisco: Jossey-Bass Inc. 512 p.

Annotation: Organizations require both management and leadership in order to survive. This book synthesizes and integrates organizational theory and research by concentrating on material useful for organizational leaders and managers. This book is divided into six sections. Section one discusses the importance of *reframing* to effective leadership and management. Additionally, it provides a brief overview of four perspectives through which issues can be viewed: *structural, human resource, political,* and *symbolic*. Section two examines the *structural frame*. The authors discuss major issues managers need to be cognizant of when designing a structural format, as well as the major difficulties managers encounter when implementing a structural design. Section three explores the *human resource perspective*, specifically, how manager behavior and organizational practices affect work force morale. Section four focuses on the *political frame*. It discusses factors that affect organizational conflict and power games including: political dynamics, scarcity, and diversity. Additionally, this section covers the difference between constructive and destructive political dynamics and identifies characteristics of a constructive politician such as agenda setting, negotiating, and building support. Section five covers the *power of symbols* to an organization, basic symbolic elements in organizations, and importance of symbols for organizations. Section six discusses the four perspectives in existing organizational and management research and theory. This book is designed to help existing and future managers identify practical and useful information from organizational theory and research.

Keywords: management, leadership, reframing, human resources, organizations, politics, morale

Deetz, Stanley A.; Tracy, Sarah J.; Simpson, Jennifer L. 2000. Leading organizations through transition. Thousand Oaks, CA: Sage. 232 p.

Annotation: This book addresses the role of communication and organizational culture during times of change, merger, transition, innovation, and globalization. Specific organizational examples are used to illustrate the processes that occur during transitional periods. The authors discuss concepts of culture, development, cultural assessment, cultural change, vision, identity, interpretation of events, ethics, management, globalization, and leadership as they are related to organizational transition. The role of communication during processes of transition within organizations is central to discussion of each of these concepts.

Keywords: organizational change, leadership, culture, cultural change

DeWine, Sue. 2001. The consultant's craft: improving organizational communication. New York: Bedford/St. Martin's. 496 p.

Annotation: DeWine offers a practical guide to consulting in a variety of organizational settings. The book has six parts, and in each part, DeWine delves into a particular consulting issue and then provides a "lesson learned" case study. First, DeWine provides a foundation for her text by defining organizational communication and the skills required of a communication manager. DeWine also explains the stages of the consultation process and how to develop credibility. In the second part of the book, DeWine addresses how to identify communication problems through collecting data and subsequently, how to develop a needs assessment. In the third part of the book, DeWine provides a collection of her training and consulting techniques, including chapters focused on technology and interventions. The fourth part of the book concerns how DeWine corrects communication failures in order to foster good listening skills, effective meetings, powerful presentations, productive teams, smooth conflict resolutions, and successful navigation of organizational change in domestic organizations and abroad. DeWine then addresses how to evaluate success. Finally, in part six, she discusses professional development opportunities for the consultant. This book provides useful diagnostic measures as well as hands-on training techniques.

Keywords: training, conflict, communication, facilitation, listening skills

Gladwell, Malcolm. 2000. The tipping point: how little things can make a big difference. Boston: Little Brown. 288 p.

Annotation: In this book, Gladwell argues that ideas, products, messages, and behaviors spread like epidemics with three underlying rules. First, they epitomize contagious behavior. Second, they show that small changes can have very big effects, and finally, these changes happen rapidly. Gladwell explains that the "tipping point" is the point in an epidemic when everything can change all at once. The book starts with a detailed explanation of the three rules of epidemics and then introduces three types of organizational members who contribute to epidemic behavior—connectors, mavens, and salesmen. Gladwell then offers a variety of different examples—from Hush Puppies to crime rates to children's TV shows—to illustrate the tipping point phenomenon. The book concludes with some practical advice for those interested in disseminating ideas. The book is written in accessible language and is especially relevant for organizational members who deal with public relations and marketing.

Keywords: public relations, tipping point

Heifetz, Ronald; Linsky, Martin. 2002. Leadership on the line: staying alive through the dangers of leading. Boston, MA: Harvard University Press. 252 p.

Annotation: The central premise of this book is that leadership is a dangerous occupation and that the adaptive work needed to overcome people's natural resistance to change can be emotionally overwhelming. This is especially the case for leaders initiating change projects. Heifetz and Linsky, two popular teachers at Harvard University's John F. Kennedy School of Government, outline key strategies for "staying alive" during the dangerous times when leadership can go awry. They label these strategies under the following general headings: *getting on the balcony* (where one steps back to get a broader perspective while staying in the game); *thinking politically* (where one monitors the quasi-equilibrium occurring with one's opponents and allies); *orchestrating the conflict* (using the stress that will naturally accumulate in the system productively); *giving the work back* (the people who are responsible for the change are given the task of accomplishing that change); and *holding steady* (where one develops the ability to stay continually focused during the heat of organizational change).

Keywords: leadership, conflict, organizational change, politics, resistance to change

Hersey, Paul; Blanchard, Kenneth H. 1977. Management of organizational behavior: utilizing human resources. Englewood Cliffs, NJ: Prentice-Hall. 343 p.

Annotation: Hersey and Blanchard discuss how managers can effect change in their organizations. The authors address behavioral motivation, the role of the leader in affecting behavior, determining effectiveness, and diagnosing the environment. Additionally, the book covers situational leadership theory, human resources, human development, and executing change. This book is useful for implementing change in organizations.

Keywords: organizational behavior, behavior, motivation, leadership, change, human resources

Kantor, David; Ober, Steven. 1999. Heroic modes: the hidden dynamics of high-stakes situations. In: Senge, Peter M; Kleiner, Art; Roberts, Charlotte; Roth, George; Ross, Rick; Smith, Bryan, eds. The dance of change: the challenges to sustaining momentum in learning organizations. New York: Currency-Doubleday: 262-269.

Annotation: In this chapter, Kantor and Ober introduce a model for dealing with conflicts that arise when employees are resistant to organizational change. Change situations are high stakes situations that cause greater fear and anxiety than low stakes situations such as everyday conflict. The authors argue that, during high stakes situations, an employee will revert to a "heroic mode," or their habitual way of dealing with emotionally charged issues. There are three heroic modes: the survivor, the fixer, and the protector. Kantor and Ober explain each mode, and tell how each mode has a corresponding light and dark aspect. When employees experience the light side of their heroic mode, they are able to persevere and excel. However, the dark side of employees' heroic modes can make them feel powerless. Learning about the heroic modes of organizational members can promote greater understanding and smoother organizational transitions.

Keywords: organizational change, conflict resolution, conflict

Kegan, Robert; Lahey, Lisa Laskow. 2001a. How the way we talk can change the way we work. San Francisco: Jossey-Bass Publishers. 256 p.

Annotation: A necessary feature of effective leadership is the ability to guide change. This book details the process of leading oneself and others toward transformational learning. Section one of this book provides a step-by-step process toward setting oneself up mentally for change. Specifically, it takes an introspective look at how internal "languages," such as those of complaint and blame, can be transformed into languages of both commitment and personal responsibility. After laying the groundwork for individual change, section two centers on how individuals can continually maintain and improve upon their ability to learn and handle change. Examples include how to transform from the "languages" of prizes and praising, rules and policies, and constructive criticism to "languages" of ongoing regard, public agreement, and deconstructive criticism. While earlier portions of this book were focused on mental languages and what individuals need for their growth, the final section discusses social languages and what leaders need to motivate change and growth in others. As a whole, this book introduces seven new languages to develop the ability to change, and once you have that ability, maintain it. Additionally, it discusses how you can use your knowledge and ability to change and to lead others.

Keywords: leadership, change, learning, rules

Klein, Stuart M. 1996. A management communication strategy for change. Journal of Organizational Change Management. 9(2): 32-46.

Annotation: Klein discusses seven communication principles that constitute a strategy to deal with problems associated with major organizational changes. The principles are: 1) message redundancy is related to message retention; 2) using several media is more effective than the using just one; 3) face-to-face communication is a preferred medium; 4) the line hierarchy is the most effective organizationally sanctioned communication channel; 5) direct supervision is the expected and most effective source of organizationally sanctioned information; 6) opinion leaders are effective changers of attitudes and opinions; and 7) personally relevant information is better retained than abstract, unfamiliar, or general information. This article is useful for those who are implementing change.

Keywords: change, communicating change, strategy, leadership

Kotter, John P. 1996. Leading change. Boston: Harvard Business School Press. 187 p.

Annotation: Kotter discusses the pressures organizations face to change in response to environmental and organizational needs. Acknowledging that major change is not simple, Kotter provides an eight stage process for creating and leading major change within organizations. Each stage coincides with errors that weaken efforts toward change. The eight sequential steps are: 1) establishing a sense of urgency; 2) creating the guiding coalition; 3) developing a vision and strategy; 4) communicating the change vision; 5) empowering broad-based action; 6) generating short-term wins; 7) consolidating gains and producing more change; and 8) anchoring new approaches in the culture. Kotter explains each step in detail to provide a method for leaders to successfully organize change.

Keywords: change, leadership, strategy

Kouzes, James M.; Posner, Barry Z. 1993. Credibility. San Francisco: Jossey-Bass Publishers. 332 p.

Annotation: According to Kouzes and Posner, the key to effective leadership is credibility, which shows as trust and confidence. This trait allows certain leaders to gain support while others fail. Based upon a combination of extensive research as well as real-world examples, Kouzes and Posner explain how leadership is both a relationship and a service. Credibility involves six disciplines: character, exploration and appreciation of others, creating and sustaining shared values, fostering capacity, developing purpose, and sustaining hope. Leaders who understand and put these disciplines into practice will further their own credibility and foster productivity in their employees.

Keywords: leadership

Larkin, T.J.; Larkin, Sandar. 1996. Reaching and changing frontline employees. Harvard Business Review. 74(3): 95-104.

Annotation: Larkin and Larkin challenge the "more is better" approach to communicating changes that need to occur among frontline employees when organizations are in trouble. The authors argue that the idea of more values, missions, visions, publications, meetings, and executive responsibility is not working. To solve communication problems, the following approach is suggested. First, communicate facts rather than values. Second, communicate face-to-face rather than via publications, videos, and large meetings. Third, to communicate with frontline employees, target frontline supervisors rather than executive management. Although this advice is contrary to ideas often associated with the charismatic executive that motivates the troops, the authors believe that in order for change to occur, it must occur with the frontline employees. Further, for frontline employees to change, they must understand the changes and receive information from trusted and familiar sources.

Keywords: change, leadership, communicating change

Lewis, Laurie K.; Hamel, Stephanie A.; Richardson, Brian K. 2001. **Communicating change to nonprofit stakeholders: models and predictors of implementers' approaches.** Management Communication Quarterly. 15(1): 5-41.

Annotation: The authors investigate how nonprofit organizations use a variety of tactics to balance relationships with stakeholders during planned organizational changes. They found six specific models that organizations use to communicate during times of change. In the first model, *equal dissemination*, organizations make sure that all stakeholders receive the same information, regardless of their status as a stakeholder. The second model, *equal participation*, differs from the first because it involves not only the dissemination of information, but also receiving feedback from the stakeholders. In the third model, *quid pro quo*, implementers give more information to those stakeholders who have something that the organization needs or wants. The fourth model, *need to know*, only informs stakeholders who really need to know, or express interest in knowing, about the changes. The fifth model, *marketing*, focuses on specific messages about the change. The sixth model, *reactionary*, is similar to crisis management communication rather than proactive change communication. However, the reactionary approach often allows for communication that is planned, whereas crisis management often deals with issues and changes as they arise. The authors note that these models are not necessarily mutually exclusive and they can be used simultaneously. The models presented here help managers understand the variety of ways in which change within the organization can be communicated to those who have specific interests in the organization.

Keywords: change, communication, stakeholders

Liedtka, Jeanne M.; Rosenblum, John W. 1996. **Shaping conversations: making strategy, managing change.** California Management Review. 39: 141-157.

Annotation: Liedtka and Rosenblum attest that management strategies are really about the management of change. Using a metaphor, they present strategy-making as a conversation with questions that lead to strategic choices: 1) What can we do? 2) What might we do? 3) What do we want to do? and 4) What ought we do? The authors add two further questions important to making strategic choices—What capabilities are we committed to developing and learning to care about? And, how can we shape tomorrow's value system to create new possibilities in partnerships with stakeholders? The creation of change by way of organizational members, management, and stakeholders together leads to a more successful change environment and the possibility of making strategy for change an ongoing and cooperative dialogue within the organization.

Keywords: change, leadership, values, stakeholders, strategy

Murgolo-Poore, Marie; Pitt, Leyland. 2001. **Intranets and employee communication: PR behind the firewall.** Journal of Communication Management. 5(3): 231-241.

Annotation: Murgolo-Poore and Pitt examine ways that information dissemination within organizations has changed, particularly with regard to intranets. Intranets allow information to be posted for employees and other authorized users. The authors explore ways in which organizations can benefit from these new knowledge-based resources, including 1) employee comfort with the technology used for information dissemination; 2) progress on the technology learning curve; 3) learning to balance technology use with human contact, and 4) preparing to change the way communication occurs. The authors suggest that if some managers and employees do not learn how to use the technology that the organization uses, important information might be lost or misunderstood.

Keywords: technology, organizational change, communication

Nutt, Paul C. 2004. **Prompting the transformation of public organizations.** Public Performance and Management Review. 27(4): 9-33.

Annotation: Nutt provides a framework for determining when a public organization is likely to be susceptible to change and offers ways for organizations to be more proactive regarding change. He also explores resistance to change and ways to cope with resistance. Resistance to change occurs when there is an imbalance between the organization's capacity for change and it's responsiveness to change. Professionally dominated, public, and routinized organizations are most resistant to change. In order to decrease this resistance, there must be an increase in the capacity for change, the responsiveness to change, or both. Various types of change (from strategic repositioning to transformation) and ways to lead change are discussed in detail. While this article proposes a research program, it is informative for leaders who are facing organizational change and/or resistance to change.

Keywords: change, organizational change, resistance to change

Nutt, Paul C.; Backoff, Robert. 1997. **Organizational transformation.** Journal of Management Inquiry. 6(3): 235-254.

Annotation: Nutt and Backoff discuss how organizations can create transformation. Using multiple approaches to transformation, the authors suggest that forces producing disequilibrium (a crisis), commitment (valuing change and investing in people), and breakout (creative leaps taken when new ideas surface that help break from the past) must be understood and managed by leaders during organizational transformation. With regard to disequilibrium, commitment, and breakout, the authors provide propositions that reflect key features of transformation, processes that can be used to produce transformation, and the stages and steps required to create a transformation. For example, to build commitment and support, Proposition 8 states that a "commitment to support change, which hinges on making investments in human resources, increases the prospect of transformation". And in response to breakouts from current practices, Proposition 11 states that "the prospect of a transformation increases when organizational leaders discontinue projects and activities rooted in current practices". In total, the authors offer 15 propositions and provide a detailed analysis of each proposition for successful organizational transformation.

Keywords: change, leadership

Repenning, Nelson P.; Sterman, John D. 2001. Nobody ever gets credit for fixing problems that never happened: creating and sustaining process improvement. California Management Review. 43(4): 64-88.

Annotation: Focusing on the entire system, this paper describes why most organizational improvement programs fail. It is not the improvement tool itself, the authors argue, but how the improvement tool interacts with the organization's physical, economic, social, and psychological structures. They show that process improvement is not just a "tool problem," but, instead, is a "systemic problem," a problem only fully understood when it is studied as an interaction between tools, workers, equipment, and managers. Using causal loop diagrams, the authors show how working harder (putting in more hours of work) does not necessarily mean working smarter (improving the company's capability). The primary causal loops described are: the *physics of improvement* (a combination of the time spent working and the capability of the process used) and the *work harder balancing loop* (performance gaps are narrowed when managers pressure employees to spend more time and energy doing work, but only for a short time). Secondary causal loops are also discussed. A considerable portion of this paper deals with the tensions created between working smarter and working harder, which leads to what the authors call the "capability trap". In this trap, managers spend most of their time improving output rather than locating the organization's systemic problems. The capability trap persists because managers often blame employees, rather than the system, for low production. The last section of this paper presents methods of overcoming the capability trap through breaking the chain of self-confirming attributions. They believe this mental model shift is absolutely necessary for a successful improvement effort. Case studies from British Petroleum and Du Pont chemical are used to support their arguments.

Keywords: systems thinking; causal loops, performance

Snow, Charles C.; Miles, Raymond E.; Coleman, Henry J., Jr. 1992. Managing 21st century network organizations. Organizational Dynamics 20(3): 5-16.

Annotation: The authors investigate how organizations are changing. They argue that network organizations have replaced the traditional multilevel, centrally managed organizational hierarchies. A network organization relies on different agencies and organizations to fulfill the organization's processes, whereas a multilevel, centrally managed hierarchy often completes the job using internal resources within the organization. The authors examine three varieties of network organizations: stable, dynamic, and internal. *Stable networks* use partial outsourcing, which can add flexibility to an organization. *Dynamic networks* outsource extensively and allow for the greatest amount of flexibility. Alternatively, *internal networks* do not engage in much outsourcing, but try to capture the commercial benefits. The authors then focus on three characteristic managerial roles that are key to successful networks: architect, lead operator, and caretaker. The *architect* facilitates operating networks without always having a specific idea of what the final product will look like. The *lead operator* connects groups together to get the job done, oftentimes building off of the work of the manager-architect. Finally, *caretakers* focus on continuing enhancement of the network process. This article pinpoints ways in which a variety of people in different agencies can work together toward a common end. It also focuses on how different styles of managers are needed to accomplish an organization's goals.

Keywords: managerial roles, organizations, networking

Van de Ven, Andrew H.; Poole, Marshall Scott. 1995. Explaining development and change in organizations. Academy of Management Review. 20(3): 510-540.

Annotation: Van de Ven and Poole review theories on the process of change in organizations and offer a new classification system. The article is written in three parts. The first part condenses existing literature on how and why organizations change into four "ideal types". These four types are life cycle, teleological, dialectical, and evolutionary. In part two, the authors argue that each ideal type only applies in certain circumstances. Part three explains how all change theories are a composite of one or more of the four ideal types. Finally, the authors discuss the strengths and limitations of existing change theory and advocate addressing multiple ideal types when researching change processes.

Keywords: organizational change, change

Weick, Karl E.; Quinn, R.E. 1999. Organizational change and development. Annual Review of Psychology. 50: 361-386.

Annotation: Examining the literature on organizational change, Weick and Quinn note that most change is conceptualized as either episodic or continuous. Episodic change is intentional, infrequent, and discontinuous, whereas continuous change is evolving and cumulative. Weick and Quinn then depict episodic and continuous change in five different ways: metaphor, analytical framework of the change process, ideal organization, intervention theory, and role of change agent. Additionally, they consider the tempo, or rhythm, of change. They characterize the sequence of episodic change as unfreeze-transition-refreeze, and the sequence of continuous change as freeze-rebalance-unfreeze. They conclude that it is important for organizational members to accept the inevitability of continuous change, and they offer practical advice for how to manage change processes through communication.

Keywords: organizational change, learning

3. What are the Challenges and Dilemmas of Leadership?

Day, David V.; Sin, Hock-Peng; Chen, Tina T. 2004. Assessing the burdens of leadership: effects of formal leadership roles on individual performance over time. Personnel Psychology. 57: 273-605.

Annotation: Day and others examine the effect of having a leader within a team on the overall performance of team members. Investigating whether changes in performance over time were impacted by individual leadership responsibilities, they studied captains of National Hockey League teams. They found that an individual's performance as part of the team increased when the player also had leadership responsibilities. The authors discuss their findings in three distinct ways. First, organizational members might be more motivated within their jobs if leadership roles are highly valued. Second, making leadership positions highly visible to others within the team might increase performance and/or motivation. Finally, a culture should be developed in which leadership responsibilities do not interfere with task performance.

Keywords: leadership, teams, culture, performance

Harvey, Jerry B. 1988. The Abilene paradox and other meditations on management. San Francisco: Jossey-Bass. 160 p.

Annotation: In this book, Harvey explores the origins and the ethical implications of moral dilemmas that face managers of American organizations. The result is a collection of everyday narratives that illustrate larger conclusions about organizational life. Chapters cover a variety of topics, including conflict resolution, alienation, and conformity. In each chapter, Harvey offers a narrative on the topic and then subsequently examines it. After examining it, he offers practical advice for how to identify that problem in an organization and either avoid or cope with it. A self-proclaimed sermonizer, Harvey uses a passionate and moralizing tone that is offset by humor. For example, an entire chapter is devoted to the metaphor of the organization as a "phrog farm," with Harvey meticulously detailing what life is like in the swamp. Overall, the book provides an entertaining anthology of parables on organizational life.

Keywords: leadership, ethics

Herzberg, Frederick. 2003. One more time: how do you motivate employees? Harvard Business Review. 81(1): 86-96. [Reprinted from 1968.]

Annotation: In this classic article, Herzberg offers practical advice on how to motivate employees. He begins by making the important distinction between "moving" and "motivating" employees. Many managers attempt to move unmotivated employees through the "kick in the ass" or KITA method. KITA can be negative and physical, such as an actual kick, or negative and psychological, which is less tangible and harder to punish. However, physical and psychological KITA can also be positive, in the form of incentives. Herzberg argues that incentives are not "pushes," but they are "pulls" from management. They get the employee to move, but do not motivate them. Employees who are moved need to be constantly reinforced, unlike those who are motivated. Herzberg then explains the motivation-hygiene theory. Environmental factors, called hygiene factors, do not cause job satisfaction, but at best cause no dissatisfaction. On the other hand, motivators, or job content factors, are what cause job satisfaction. This theory is then applied to a case study. Herzberg offers practical suggestions for implementing the motivation-hygiene theory through vertical loading, which means adding motivation factors to an employee's job.

Keywords: leadership, motivation

Iskat, Gregory J.; Liebowitz, Jay. 2003. What to do when employees resist change. Supervision. 64(8): 12-15.

Annotation: The authors discuss thirteen steps that managers can take to ensure employee commitment when implementing organizational change. These steps can help managers avoid employee resistance during times of transition. The authors organize the thirteen steps according to three stages of organizational change—unfreezing, making the transition, and refreezing. In the unfreezing stage, managers open employees up to the possibility that there are alternative ways of doing things. In the transition change, managers gradually implement changes. In the refreezing stage, policies and human resource systems are changed in order to reinforce the new way. The authors go on to explain, in detail, each step that can be taken during each stage of transition to decrease employee resistance and ensure a smooth transition.

Keywords: organizational change, resistance to change, leadership

Kegan, Robery; Lahey, Lisa Laskow. 2001b. The real reason people won't change. Harvard Business Review. November: 84-92.

Annotation: Kegan and Lahey examine employee reactions to organizational change. They argue that many employees are devoted to a "competing commitment" that could be misconstrued as a personal resistance to change. In order to facilitate smooth organizational transitions, Kegan and Lahey explain what competing commitments are, and give several real-life examples. They outline how to help employees overcome competing commitments. First, a guided group discussion can assist in uncovering the competing commitment and its underlying assumptions. Once the commitment is discovered, a five-step process can be used to help employees cope with inner conflicts. This article is a practical guide to how managers can thoughtfully work with employees who seem inexplicably resistant to change.

Keywords: organizational change, resistance to change, conflict

Nutt, Paul C.; Backoff, Robert W. 1992. Strategic management of public and third sector organizations. San Francisco: Jossey-Bass Publishers. 512 p.

Annotation: In efficient organizations, strategic management is an ongoing process. In order to implement this process, leaders and managers need to understand the strategic management process and be able to tailor the process to their organization. To that end, this book is organized into five sections. Section one discusses strategy and the challenges associated with it, including recommended responses to such challenges. Section two differentiates between public, third sector, and private organizations. This section discusses the unique challenges associated with strategy in each sector. Section three overviews the strategic management process including party roles. It includes a section on how to tailor the strategic management process to each unique organization. Section four covers techniques that can be used in each step of the strategic management process. These include techniques for prioritizing information and how to select techniques to form a specific strategy. Section five provides real-life examples. This book may be useful for managers and leaders who desire a basic understanding of strategic management. Additionally, those wishing to tailor the strategic management process to their organization may also find this book useful.

Keywords: leadership, strategy

Quick, James C.; Quick, Jonathan D. 1984. Organizational stress and preventive management. New York: McGraw Hill. 346 p.

Annotation: In this book, Quick and Quick argue that, although mismanaged stress can lead to poor quality productivity and management labor conflicts, stress is also essential for growth, change, development, and performance. Stress is both beneficial and destructive, and the authors write the book with a concern for both individual and organizational health. They emphasize the diagnosis of stress, philosophy of preventative management, and practice of preventative management at both the organizational and individual level. Organizational-level preventative management includes task redesign, participation management, flexible schedules, career development, physical setting design, role analysis, goal setting, social support, and team building. Individual-level preventative management includes managing perceptions of stress, managing the work environment,

lifestyle management, relaxation training, physical outlets, emotional outlets, counseling, psychotherapy, and medical care. The authors state that destructive consequences of stress can be avoided through application of the prescribed preventative management strategies discussed.

Keywords: stress

Strebel, Paul. 1996. Why do employees resist change? Harvard Business Review. 74(3): 86-92.

Annotation: Strebel focuses on the fact that managers and employees who view change in different ways may prevent change management from working. Given that, he suggests a personal contract between the employee and organization that has three specific focus areas: *formal, psychological,* and *social.* The *formal* aspect focuses on the basic tasks/requirements of the job (for example, job description). The *psychological* aspect delineates typically implicit expectations and obligations. Finally, the *social* aspect explains the organizational values and the ways that those are played out in the organization. To assist with the process, Strebel provides possible questions that can be considered within each section. In conclusion, Strebel gives examples of the use of personal contracts in organizational life.

Keywords: change, leadership, organizational behavior, resistance to change

E. Organizational Learning

The science of how organizations, and the people within those organizations, learn is a relatively new field of study (Senge 1990). Organizational learning implies not only the acquisition of knowledge, but the transfer of that knowledge to employees within the organization. In this section, we provide an introduction to the basic concepts of organizational learning (Argyris 1990; Kofman and Senge 1993; Senge 1990), including a discussion of the business concept of "best practices" (Camp 1989). In addition to providing a theoretical background for the science of organizational learning, we provide further citations that will help guide managers through the creation of a learning organization on their work unit (Garvin 2000; Goleman 2000; Schwarz and others 2005; Senge and others 1994). Also included here are studies of how organizations learn in high hazard work environments (Carroll 1998; Kantor and Ober 1999).

Author's Picks

- **Argyris, Chris. 1990. Overcoming organizational defenses: facilitating organizational learning.** Needham Heights, MA: Allyn and Bacon. 180 p.
- **Garvin, David A. 2000. Learning in action: a guide to putting the learning organization to work.** Boston: Harvard Business School Press. 272 p.

- Senge, Peter M. 1990. **The fifth discipline.** New York: Doubleday. 432 p.
- Senge, Peter; Kleiner, Art; Roberts, Charlotte; Ross, Richard; Smith, Bryan. 1994. **The fifth discipline fieldbook: strategies and tools for building a learning organization.** New York: Doubleday/Currency. 594 p.

1. What is Organizational Learning?

Argyris, Chris. 1990. Overcoming organizational defenses: facilitating organizational learning. Needham Heights, MA: Allyn and Bacon. 180 p.

Annotation: Organizations often suffer because workers and managers avoid embarrassment by turning a blind eye toward mistakes. This book focuses on errors that are consciously buried in order to avoid situations that might damage individual reputations. Using examples from businesses across the United States and government errors like the Challenger disaster, Argyris highlights the processes that produce and generate error in organizations. He then offers solutions to reduce error by minimizing the tendency of organizations to become defensive about errors that make them or their managers look bad. Lastly, Argyris addresses how to apply these solutions in actual practice. This book has direct application for managers interested in understanding how errors develop from organizationally defensive reactions and how to resolve them in productive ways.

Keywords: organizations, organizational learning, Challenger, image, accidents

Kofman, Fred; Senge, Peter. 1993. Communities of commitment: the heart of learning organizations. Organizational Dynamics. 22(2): 5-23.

Annotation: In this article, Kofman and Senge explore shifts in contemporary management principles. They argue that the common organizational complaints of reactiveness, competition, and fragmentation are larger societal issues. These problems cannot simply be solved, but need to be dissolved through a new way of thinking, feeling, and being. Through what they call the "Galilean Shift," *reactiveness* can become *creating*, *competition* can become *cooperation*, and *fragmentary thinking* can become *systemic*. Such a shift is impossible without personal commitment to organizational change and community. Traditional analytical models involve a three part process: 1) break the system into parts, 2) study each isolated part, and 3) understand the whole through understanding the parts. In comparison, Kofman and Senge's Galilean Shift also has three parts: 1) the primacy of the whole, 2) the community nature of self, and 3) language as generative practice. From these three assumptions, the authors discuss several operating principles and advocate community collaboration in order to devise new and inspirational models of organization.

Keywords: learning organization, community, systems thinking, Galilean Shift

Senge, Peter M. 1990. The fifth discipline. New York: Doubleday. 432 p.

Annotation: This book takes a systems approach to organizational learning. Senge argues that "a learning organization is a place where people are continually discovering how they create reality. And how they can change it". This book explains why a learning organization is important and describes how to create one. The book is divided into five sections that address how individuals create their reality, as well as how they can change it. Section two delineates how the fifth discipline serves as a cornerstone for a learning organization. Section three discusses the bases of creating a learning organization including personal mastery, mental imagery, shared goals and visions, and team learning. Section four discusses obstacles that interfere with forming a learning organization, as well as ways to assist in organizational member learning. Section five suggests an additional discipline to account for future developments. This book will aid organizational members involved in establishing an atmosphere conducive to learning. It also discusses the role of management in establishing a culture of learning.

Keywords: systems theory, organizational learning

2. How are Learning Organizations Created?

Camp, Robert C. 1989. Benchmarking: the search for industry best practices that lead to superior performance. Milwaukee, WI: ASQ Quality Press. 299 p.

Annotation: Organizations need reliable ways to adjust business practices when work conditions change. A proven method of successfully learning and applying organizational best practices is through a process called *benchmarking*. This book introduces the benchmarking process by showing how to: 1) structure and conduct investigations, 2) analyze and measure current practices, as well as predict future changes in those practices, and 3) apply an action plan to put lessons learned into practice.

Keywords: benchmarking, learning, performance

Carroll, John S. 1998. Organizational learning activities in high-hazard industries: the logics underlying self-analysis. Journal of Management Studies. 35: 699-717.

Annotation: Carroll begins by discussing how different staff members in an organization know different things about how work is accomplished. For an organization to run properly, these staff members must engage in organizational learning, which means facilitating the development of organizational knowledge by supporting each other and acknowledging staff members interdependent nature. Carroll investigates how organizational learning occurs by studying two high hazard industries: nuclear power plants and chemical processing plants. The results indicate that, given the time constraints on staff, problems are often initially attributed to equipment or process. It is only with careful investigation that cultural and communicative problems are uncovered. For example, staff members often view other departments as living in different "thought worlds". Carroll separates these thought worlds into four types: abstract, concrete, anticipation, and resilience. He argues that incident review processes should be more comprehensive and take all of these types into account in order to fix problems and facilitate greater organizational learning.

Keywords: organizational learning, risk, uncertainty, culture, incident review

Garvin, David A. 2000. Learning in action: a guide to putting the learning organization to work. Boston: Harvard Business School Press. 272 p.

Annotation: Creating a learning organization is a much needed, yet difficult to implement asset for any organization. While most managers agree on the importance of building this skill, few understand how to get started, what tools and techniques are needed, or even how to tell when they have truly created a learning organization. This book begins by describing the basic characteristics and processes of a learning organization. Tests to gauge organizational progress are provided as well as practical tips on how to raise the priority placed on learning. Suggestions are also provided on how to overcome common impediments to learning. This book then covers various learning modes and processes. While each involves the same fundamental process of acquiring, interpreting, and applying information, each mode applies them differently. For each mode of learning, Garvin discusses basic properties and necessary supporting conditions and steps for success. Garvin concludes by concentrating on the role of the manager in building a learning organization. Common challenges managers face are discussed as well as ways managers and executives can become better leaders themselves. By focusing on the importance of practice and attention to detail, this book is a useful guide for those who wish to build a learning organization but are unsure how to get started.

Keywords: organizational learning, management

Goleman, Daniel. 2000. Working with emotional intelligence. New York: Bantam. 400 p.

Annotation: Based on his earlier book, "Emotional Intelligence," Goleman applies years of research to this practical guide on emotional intelligence in organizations. In the first part of the book, Goleman makes a "hard case for soft skills" by arguing that emotional intelligence is paramount to IQ, or expertise, in determining who will excel at a job. In the second part of the book, Goleman explores the crucial elements of self-mastery and links these to specific job capabilities. The third part of the book moves from self-mastery to interpersonal skills that help employees perform. Many skills are covered, including empathy, listening skills, collaboration, teamwork, and networking. In part four, Goleman offers well-researched guidelines for improving emotional intelligence capabilities. These guidelines include caveats and best practice models. Finally, part five details a case study in an emotionally intelligent organization and explains how such an organization is able to weather turbulent times.

Keywords: high reliability organizing, emotional intelligence, management, listening skills, networking

Kantor, David; Ober, Steven. 1999. Heroic modes: The hidden dynamics of high-stakes situations. In: Senge, Peter M; Kleiner, Art; Roberts, Charlotte; Roth, George; Ross, Rick; Smith, Bryan, eds. The dance of change: the challenges to sustaining momentum in learning organizations. New York: Currency-Doubleday: 262-269.

Annotation: See annotation in Section II.D.2. *How do leaders create/manage change?* p. 33.

Langer, Ellen J. 2005. On becoming an artist: reinventing yourself through mindful creativity. New York: Random House. 288 p.

Annotation: In this book, Langer augments her previous work on mindfulness and artistic nature with insights on creativity. Langer discusses how creativity is not a rare trait, but a part of every person's makeup. While all people have the ability to express themselves creatively, many people undervalue themselves, which serves to undermine expressions of creativity. By discussing a number of her own experiments and those of her colleagues, Langer discusses mindfulness and its relation to creativity. Through these experiments, a number of insights are revealed, such as benefits of uncertainty and how people learn to ignore differences.

Keywords: mindfulness, uncertainty

Langer, Ellen J. 1997. The power of mindful learning. Reading, MA: Addison Wesley. 167 p.

Annotation: Mindful learning takes place with an awareness of contexts and of the ever-changing nature of information. A mindful approach is distinguished by three characteristics: the continuous creation of new categories, openness to new information, and an implicit awareness of more than one perspective. Learning is hindered by popular misconceptions about learning. These myths are, first, that the basics must be learned to such an extent that they become second nature. Second, paying attention means staying focused on one thing at a time. Third, delaying gratification is necessary and important. Fourth, rote memorization is essential to education. Fifth, forgetting is a problem. Sixth, intelligence is knowing "what's out there". Seventh, there are right and wrong answers. This book explains how these myths undermine true learning. Langer applies the theory of mindfulness to multiple contexts and explains how mindfulness or conditional learning is effective and how, alternatively, learning without this awareness both limits learning and can be a precursor to failure.

Keywords: mindfulness, learning

Leonard, Dorothy; Swap, Walter. 2005. Deep smarts: how to cultivate and transfer enduring business wisdom. Boston: Harvard Business School Press. 288 p.

Annotation: This book describes a process of extracting knowledge from inside the heads of people who possess high expertise in their work, and then describes numerous ways this knowledge can be delivered to people with less expertise. People with high expertise are said to have "deep smarts" about their work, a form of work-related knowledge that is both explicit and tacit. Their peers may know they are good at what they do, but the practitioners themselves may not know exactly how they put the whole package of their work-related expertise together. Employees who have deep smarts have many attributes, including the ability for swift decision making, to find solutions to novel situations, to ascertain when rules don't apply, and to recognize patterns. But knowledge locked in people's heads doesn't help improve the performance of organizations—this knowledge must be cultivated and passed on to employees who are still working. Leonard and Swap discuss coaching as an excellent method of knowledge transfer. In an age when many organizations' most experienced people are retiring, the guidelines for knowledge capture and transfer outlined in this book deserve close attention.

Keywords: knowledge transfer, coaching

Schwarz, Roger; Davidson, Anne; Carlson, Peg; McKinney, Sue. 2005. The skilled facilitator fieldbook: tips, tools and tested methods for consultants, facilitators, managers, trainers and coaches. San Francisco: Jossey-Bass. 576 p.

Annotation: See annotation in Section III.C.2. *What is effective crisis communication?* p. 65.

Senge, Peter; Kleiner, Art; Roberts, Charlotte; Ross, Richard; Smith, Bryan. 1994. The fifth discipline fieldbook: strategies and tools for building a learning organization. New York: Doubleday/Currency. 594 p.

Annotation: This fieldbook, though dated, remains an excellent resource for anyone interested in using the principles of learning organizations in a field setting. It is a companion to Senge's book, "The Fifth Discipline: The Art and Practice of the Learning Organization". The fieldbook is a series of notes, reflections, and exercises written by people who have "been in the field" using and learning how to apply Senge's five principles of a learning organization. The five principles—personal mastery, mental models, shared vision, team learning, and systems thinking—are used as chapter headings, and over 60 individual authors suggest actions one might take to develop a deeper ability to learn and grow, both individually and organizationally. Each of the five sections begins with a succinct definition and a brief overview of the learning organization principle being described. Within each section, strategies, tools, and notes for further reading are presented to help the practitioner more fully develop skills in using that principle. Numerous sidebars with pictures graphically display how these tools can be used. Some of the more fully developed concepts include developing feedback loop charts to be used while dealing with systems thinking problems; telling a good story to induce learning; strategies for working with mental models and the conflict that can occur when mental models clash; designing a dialogue discussion group; facilitating a meeting using the tools ladder of inference; and balancing advocacy and inquiry. The final chapter deals with the "frontiers" of the learning organization—organizations as communities, the core processes of organizations, organizations as natural learning laboratories, and using flight simulators to model management problems.

Keywords: organizational learning,

Weick, Karl E. 1998. Improvisation as a mindset for organizational analysis. Organizational Science. 9(5): 543-555.

Annotation: Weick uses the metaphor of jazz improvisation to discuss the way people verbally communicate about organizational improvisation. He argues that the descriptions and processes used to discuss composing on the spur of the moment provide a lexicon for discussing concepts central to organizational theory. Weick places an emphasis on the holistic nature of what occurs when people improvise. Specifically, this article examines various aspects of improvisation, such as degrees of improvisation, forms of improvisation, and cognition in improvisation. Weick ends with a list of 13 characteristics of groups with a high capability to improvise and comments about implications for theory and practice.

Keywords: improvisation, cognitive processes, communication, groups

Wilber, Ken. 2000. A theory of everything: an integral vision for business, politics, science and spirituality. Boston: Shambhala. 189 p.

Annotation: Wilber provides a brief overview of a "Theory of Everything," which attempts to include matter, body, mind, soul, and spirit as they appear in self, culture, and nature. This theory provides an integral vision, which is a more holistic approach to the world in which we live. In order to explain the theory, Wilber devotes the first four chapters to discussing the characteristics of an integral model. Characteristics include the acceptance of multiple levels, or "waves," of existence, numerous streams of development, multiple stages of consciousness, and the influential forces of the social system. The next three chapters are devoted to the theory's importance in the real world in realms such as politics, governance, medicine, business, and education. Wilber concludes with a chapter on taking the theory of everything into everyday life through an "integral transformative practice".

Keywords: Theory of Everything, politics

F. Team and Crew Dynamics

Teams are groups of two or more people who have the same performance objectives or goals. There are many functions and processes that affect teams, and assessing the characteristics of effective teams helps uncover strategies for working with and employing teams. This section first addresses what teams are, how they are becoming increasingly important (Katzenbach and Smith 1993a), and how different types of teams work toward different goals. Crews are described as a specific type of work team (Webber and Klimoski 2004).

Next, this section offers practical tips for managing teams (Scholtes and others 2003) and uses examples from different organizations to illustrate the many goals and characteristics of teams that make them more efficient and effective (Larson and LaFasto 1989). Different types of teams include self-directed and leader-directed teams. Self-directed teams are small, functional units that are structured to manage their own affairs (Holpp 1993). They are different from leader-directed teams because they require effective external leadership (Manz and Sims 1987), which contrasts with leader-directed teams. Although self-directed teams can reduce the cost of middle management and increase productivity, self-directed teams are challenging to implement because they reorganize leadership roles and require additional training (Holpp 1993). A clear understanding of teams is essential for organizations that are considering moving toward using leader-based or self-directed teams.

Author's Picks

- Barker, James R. 1993. **Tightening the iron cage: concertive control in self-managing teams.** Administrative Science Quarterly. 38(3): 408-437.

- Driessen, Jon. 2002. **Crew cohesion, wildland fire transition, and fatalities.** Missoula, MT: United States Department of Agriculture, Forest Service, Technology and Development Program, TE02P16—Fire and Aviation Management Technical Services. 15 p. Available: http://fsweb.mtdc.wo.fs.fed.us. [July 6, 2007].

- Katzenbach, Jon R.; Smith, Douglas K. 1993. **The wisdom of teams: creating the high-performance organization.** New York: Harper Business. 352 p.

- Scholtes, Peter R.; Joiner, Brian L.; Streibel, Barbara J. 2003. **The team handbook.** 3rd ed. Joiner/Oriel Inc. 400 p. Spiral bound.

1. What are Teams/Crews?

Driessen, Jon J. 1990. The supervisor and the work crew. Missoula, MT: U.S. Department of Agriculture, Forest Service, Missoula Technology and Development Center.

Annotation: This short publication serves as a training course for new supervisors in the USDA Forest Service. The course is broken into five sessions. In the first session, Driessen discusses the qualities and skills of good supervisors. Supervisors must fully understand the concept of a crew, and then make a good impression on their own crew. Good supervisors must also have authority, recognize the importance of crew cooperation, and involve their crews in decision making processes. The second session of this course covers arbitration, setting expectations, and training of crews. Session three addresses team building and new crew member integration. In session four, Driessen stresses safety and health, including a broad range of strategies from open communication to selection and maintenance of equipment. Finally, the training course ends with a session on overall physical fitness and crew harmony. The Forest Service publication is a practical guide to how to manage the day-to-day responsibilities of an effective supervisor.

Keywords: leadership, management, teams, team building, conflict, training

Larson, Carl E.; LaFasto, Frank M.J. 1989. Teamwork: what must go right/what can go wrong. Newbury Park, CA: Sage. 152 p.

Annotation: The authors begin by providing a broad definition of teams: a team has two or more people, it has a specific performance objective to attain, and coordination of activity among the members of the team is required for the attainment of the team's objective. The article then investigates goals of teams, team structures, team member competency, commitment, collaboration, performance, support, and leadership. Larson and LaFasto propose eight characteristics that are similar across a broad range of teams and then use these characteristics to explain how and why effective teams develop. Finally, after identifying factors for effective team development, the authors outline performance measures for monitoring and providing feedback to teams.

Keywords: teams, team leadership, performance

Lipnack, Jessica; Stamps, Jeffrey. 1997. Virtual teams: reaching across space, time, and organizations with technology. New York: John Wiley and Sons. 288 p.

Annotation: Lipnack and Stamps focus on the emerging and transforming role of the team in organizations. With the increasing role of technology in the workplace, the team has transformed from a large group of people working in proximity to a small group of people who, through technology, span boundaries. This relatively new position of the team brings forth a new set of benefits and challenges for organizations. The authors discuss the characteristics of this new type of team through a combination of personal insights, case studies, and theory. Managers who are struggling with the increasing role of technology in their organization may find this book useful. Additionally, this book is useful for providing a basic overview of virtual teams and their role in the workplace.

Keywords: technology, teams, networking

Webber, Sheila Simsarian; Klimoski, Richard J. 2004. Crews: a distinct type of work team. Journal of Business and Psychology. 1(3): 261-279.

Annotation: In this study, Webber and Klimoski introduce the Crew Classification Scale (CCS), a scale by which to measure a work team's "crewness". Their purpose is to provide empirical proof that crews (such as firefighters in the Forest Service organization) are a distinct type of work team that should be researched independently of other team "types". They describe work crews as typically short-lived groups with clear role/position assignments where repetition of specific actions and/or responses is built into the job. For example, a work crew may be instructed to form a line on a fire. Since their duty is to build a line, and this requires them to make certain cuts in the land and work together in certain ways, their work reinforces a set of repetitious behaviors that govern their actions as long as they are working together as a fire line crew. Relations among crew members are defined by their function in the crew, where one member can be easily replaced by another person of equal skill or function, and where technology, such as tools, hardware, and the procedures that dictate their use, is considered the most important and binding element. Role and status relations often restrict upward mobility. Managers and practitioners can use the scale to more systematically identify work crews from other types of work teams and thereby increase the specific knowledge related to crew behavior.

Keywords: crews, teams

2. How Do Teams/Crews Work?

Barker, James R. 1993. Tightening the iron cage: concertive control in self-managing teams. Administrative Science Quarterly. 38(3): 408-437.

Annotation: Self-managing teams are characterized by strong identification with team-approved values, peer-enforced behavior, and the ability to quickly adapt to their environments. This article takes a close look at how self-managing teams function and develop over time. There are three phases of team development. First, team members begin by establishing behavioral actions that support larger organizational values. Second, as teams work together and solve problems, they begin to form and accept unspoken values that indirectly control their actions. When situations arise that demand a decision, they act as the team would have them act by relying upon the team values to guide

behavior. Third, teams formalize unspoken values and beliefs so that new members can understand how to behave within the team. This has direct application for managers who are looking to understand how teams work, as well as consider the strengths and weaknesses of self-directed teams.

Keywords: identification, teams, decision making

Driessen, Jon. 2002. Crew cohesion, wildland fire transition, and fatalities. Missoula, MT: United States Department of Agriculture, Forest Service, Technology and Development Program, TE02P16—Fire and Aviation Management Technical Services. 15 p. Available: http://fsweb.mtdc.wo.fs.fed.us. [July 6, 2007].

Annotation: Crew cohesion was seen as an important dynamic that contributed to the tragedies at the Mann Gulch, South Canyon, and Thirtymile wildfires. This paper distinguishes the need for two types of crew cohesion to ensure safety and efficiency on wildfires: 1) intracrew cohesion and 2) intercrew cohesion. Intracrew cohesion describes how a single crew, such as a smokejumper crew, is individually bonded together. Intercrew cohesion describes how multiple crews, such as hotshots, smokejumpers, and BLM crews who are used to fight large wildfires, are bonded. This paper first looks at how crew cohesion contributed to the Mann Gulch, South Canyon, and Thirtymile wildfires. It then explores how social science research has studied cohesion, and it makes a case for the importance of crew cohesion on dangerous transition fires. Since a majority of fire fatalities occur on transition fires, more research needs to focus on identifying the stages and types of transition fires and how to increase crew cohesion during such critical situations. This article offers practical training suggestions for fire managers to improve crew cohesion during transition wildfires.

Keywords: crew cohesion, crews, Mann Gulch, South Canyon, Thirtymile, training

Katzenbach, Jon R.; Smith, Douglas K. 1993a. The discipline of teams. Harvard Business Review. March-April: 111-120.

Annotation: This article is a brief summary of Katzenberg and Smith's book, "The Wisdom of Teams" (see next annotation). The authors believe that teams will become the primary unit of performance in high-performance organizations and teams will enhance existing structures. They state that top management must recognize a team's ability to deliver results, deploy teams strategically when they are the best tool for the job, and discipline teams in a way that will make them most effective. Disciplining teams is a multi-faceted task. First, the combination of purpose and specific goals are essential to performance. The right team size is important, with 2 to 25 being the most effective range. Teams must develop a mix of skills, including technical or functional expertise, problem solving and decision making skills, and interpersonal skills. Teams also need a common approach. With a common purpose, approach, and goals, mutual accountability grows and accomplishment is most effective. Top management must assess team types (teams that recommend things, make or do things, and those that run things) in order to discipline and employ them effectively.

Keywords: teams, performance

Katzenbach, Jon R.; Smith, Douglas K. 1993b. The wisdom of teams: creating the high-performance organization. New York: Harper Business. 352 p.

Annotation: The authors of this book explore research findings about teams that represent both common and uncommon sense. Their research findings come from stories of a variety of teams with wide-ranging performance records. Common sense findings include the following: a demanding performance challenge tends to create a team, the application of basic team rules is often overlooked, team performance opportunities exist in all parts of the organization, teams at the top are the most difficult, and most organizations intrinsically prefer individual over group (team) accountability. Uncommon sense findings include: companies with strong performance standards seem to spawn more "real teams" than companies that promote teams per se, high-performance teams are extremely rare, teams work well in hierarchical organizations, teams naturally integrate performance and learning, and teams are the primary unit of performance in many organizations. Using case studies of actual teams, the authors explore the obstacles that teams face, team performance, and team leadership.

Keywords: teams, team leadership, performance, rules

3. How Do Managers Lead Teams/Crews?

Driessen, Jon J.; Haubenreiser, Jennifer. 1996. Making a crew. Part I: putting a crew together. Part II: keeping a crew together. Missoula, MT: U.S. Department of Agriculture, Forest Service, Missoula Technology and Development Center. (These manuals and videos is available at the Missoula Technology and Development Center; 5785 Hwy 10 W; Missoula, MT 59808; Phone: 406-329-3900).

Annotation: This two part course includes two manuals and two short instructional videos written for leaders who need to acclimate new Forest Service crew members and to recharge seasoned workers. The authors begin by educating the reader and potential facilitator about how to help students get the most out of the course. Advice includes: be prepared, focus on main issues, encourage class discussion, and use narrative examples. There are six sessions in the first part of the course: work skills, workers, teamwork, supervisors, harmony, and a summary session. There are four sessions in the second part of the course: the nature of work,

individual problems, interpersonal problems, and changes in the Forest Service (when this course was developed). Each session includes main points that facilitators should emphasize, a video tie-in, and discussion questions. The course is specifically designed to educate Forest Service crew members and includes developing general team building skills.

Keywords: crews, teams, team building, conflict

Holpp, Lawrence. 1993. Self-directed work teams are great, but they're not easy. Journal for Quality and Participation. December: 64-70.

Annotation: The author of this article provides accessible diagrams that show different aspects of self-directed teams. Self-directed teams are defined as small, functional units that are structured to manage their own affairs. Throughout the article, explanations, diagrams, and tables illustrate: 1) steps for moving from a leader-based team to a self-directed team; 2) examples of self-directed teams in action; 3) sequences of training meetings and topics for self-directed teams; 4) progression of empowerment and responsibilities in self-directed teams; 5) needed teamwork, problem solving, and implementation skills; 6) suggested training modules; and 7) managerial role changes and results. They provide dos and don'ts for managers attempting to employ self-directed teams. The authors make it clear that self-directed teams can be very effective, but they are not easy, and change toward self-directed teams will not happen overnight.

Keywords: teams, self-directed teams, team leadership, training

Manz, Charles C.; Sims, Henry P. 1987. Leading workers to lead themselves: the external leadership of self-managing work teams. Administrative Science Quarterly. 32:106-128.

Annotation: Manz and Sims begin by posing the question: If self-managing teams are truly self-managing, then why should an external leader be required? To shed light on this question, the authors identified behaviors of self-management-team leaders that are effective in encouraging self-management. These behaviors include facilitation through self-observation, self-evaluation, and self-reinforcement. In other words, external leaders of self-managing teams reinforce the importance for team members to observe, evaluate, and reinforce each other instead of depending on outside assistance. The external leaders of self-managing teams differed from traditional leaders in their commitment to a bottom-up rather than top-down hierarchy. These leaders encourage team member independence and act in a more consultative role. Manz and Sims conclude that, although different from traditional and participative leaders, such leaders play a legitimate role in team success.

Keywords: teams, self-managing teams, leadership

Scholtes, Peter R.; Joiner, Brian L.; Streibel, Barbara J. 2003. The Team Handbook. 3rd ed. Joiner/Oriel Inc. 400 p. Spiral bound.

Annotation: This is a popular (over a million copies sold of all editions in print), practical "how to" manual on getting high quality work within team settings. The approach used in this book is based on the work of management consultant, Dr. W. Edwards Deeming, who was one of the early leaders in introducing total quality management to industrialized countries. The original premise of the book discussed in earlier editions still holds true: project teams need more than technical knowledge to successfully manage projects and ensure quality, efficiency, and low-costs. They also need to develop skills in team building, planning, conducting productive meetings, managing logistics, gathering and analyzing data, communicating information, and finally, implementing the project. Regardless of the topic being discussed, the team-concept is what holds each chapter of this book together. This third edition includes updated material on how to start quality improvement initiatives; dealing with conflict; advanced tools and techniques; the roles of team leaders, coaches, sponsors and team members; using design experiments to identify and control process variation; force field analysis; completing team assignments virtually; and managing project pipelines. This spiral bound book contains numerous sidebars, worksheets, and step-by-step instructions.

Keywords: teams, team building

III. Understanding Organizations in High Risk Contexts

This section addresses issues, challenges, and processes that impact organizational members who work in high risk or highly uncertain environments. For the wildland firefighting community, the readings included here provide insight into organizational challenges that may be unique to organizations like firefighting crews. While many of the authors included in this section cover processes such as decision making, which are summarized in previous sections, these authors focus on the nuances and complexities that make management of the unexpected both unique and difficult. Readings annotated in this section provide a comprehensive overview of both foundational and state-of-the-art research on human factors in high risk or uncertain contexts. Three main topics are explored. The first subsection examines the growing body of research that focuses on how organizations successfully manage risk and uncertainty. The second subsection focuses specifically on "high reliability organizations" and provides examples of lessons learned from organizations that operate with few errors under difficult circumstances. The third subsection surveys resources related to preparing for and managing public relations challenges associated with crisis.

A. Risk and Uncertainty

Wildland firefighters deal with risky and uncertain situations on a daily basis. Readings in the first section comment on how social trends largely shape what people consider as "risky" (Douglas and Wildavsky 1982; MacLean and Mills 1990; Popkin 1990). The second section uses case studies, including military (Cohen and Gooch 1990; Snook 2000), space shuttle (Chiles 2002; Perrow 1999; Starbuck and Milliken 1988; Tompkins 2005; Vaughan 1996), air (Weick 1991), and 9/11 (National Commission on Terrorist Attacks Upon the United States 2004) disasters to better understand the causes of disaster. The final section is dedicated to the practical application of risk management strategies taken from a variety of high risk organizations including wildland firefighting (Alder 1997), aviation (Tompkins 1990; Krieger 2005), space (Columbia Accident Review Board 2003; Tompkins 1993), and energy (Hopkins 1999; Hopkins 2000). Broader discussions of how organizations can learn from, prevent, and manage uncertainty are provided by Argyris (1990), Carroll (1998), Cleaves and Haynes (1999), Kahneman and others (1996), Lagadec (1993),

Leveson (2004), Perrow (2007), Reason (1997), Staw and others (1981), and Weick and Sutcliff (2007). The readings in this section include hands-on techniques and practical advice essential for managing uncertain situations.

Author's Picks

- Kahneman, Daniel; Slovic, Paul; Tversky, Amos, eds. 1986. Judgment under uncertainty: heuristics and biases. New York: Cambridge University Press. 544 p.
- Perrow, Charles. 1999. Normal accidents: living with high-risk technologies. Princeton, NJ: Princeton University Press. 386 p.
- Reason, James. 1997. Managing the risks of organizational accidents. Brookfield, VT: Ashgate. 252 p.
- Tompkins, Phillip K. 2005. Apollo, Challenger, Columbia: the decline of the space program. Los Angeles: Roxbury. 288 p.
- Vaughan, Diane. 1996. The Challenger launch decision: Risky technology, culture, and deviance at NASA. Chicago, IL: The University of Chicago Press. 592 p.

1. How Does Society Define Risk?

Douglas, Mary; Wildavsky, Aaron. 1982. Risk and culture: an essay on the selection of technological and environmental dangers. Los Angeles: University of California Press. 224 p.

Annotation: Perceptions of risk and danger are largely based on what society and organizations choose to value and what they choose to fear. Often, these values are shaped by larger organizational and political interests that bias perceptions of risk, danger, and responses that reinforce the values they match. Scientific disagreement on what constitutes acceptable human risk when dealing with issues of human health, technology, and the environment support the idea that risk and risk aversion are based on collective social values and are only selective representations of danger. Managers who trust completely in society or the organization to define what is safe and what is risky may not see actual dangers until it is too late.

Keywords: risk, culture, technology, politics

MacLean, Douglas; Mills, Claudia. 1990. Conservatism, efficiency, and the value of life. In: Kirby, Andrew, ed. Nothing to fear: risks and hazards in American society. Tucson, AZ: University of Arizona Press: 53-74.

Annotation: Many organizations are faced with unavoidable dangers that may harm employees during their regular work. However, despite organizational efforts, it is often unrealistic for employees to be responsible for avoiding all possible risk. Employees must get the job done despite the dangers they may face. In this chapter, MacLean and Mills discuss the arguments for and against a conservative view of risk taking and then argue in favor of a "bounded" notion of conservative behavior. Since society largely determines what is considered "risky," managers need to be aware of three structural levels that control how organizations define "risk". First, risk is defined by those who make the laws for and against taking risks on a national and state scale. Second, risk is reshaped by those who interpret the previously defined laws in order to apply them to the organization. Third, individuals and agencies that must assess organizational risk may change the bias of risk taking in favor of a more conservative approach.

Keywords: risk

Popkin, Roy S. 1990. The history and politics of disaster management in the United States. In: Kirby, Andrew, ed. Nothing to fear: risks and hazards in American society. Tucson, AZ: University of Arizona Press: 101-130.

Annotation: This chapter overviews the history of hazard management in the United States, and defines what constitutes a "disaster" by federal standards. Popkin provides a history of federal and state policies and programs established and maintained in the United States since 1803. Despite historic moves to ensure federally funded hazard protection, Popkin argues that there is always a need for political pressure on federal representatives to ensure future legislative support. Understanding the history of disaster management in the United States can help managers see their place within the current federal structure. It also emphasizes the need for political savvy and persistence to ensure funding for current and future disaster management initiatives.

Keywords: disaster, politics

Wildavsky, Aaron. 1987. Searching for safety. New Brunswick, NJ: Transaction. 253 p.

Annotation: Safety and danger are usually seen as opposites that exclude one another in practice. Safety is often seen as the absence of danger, and danger is often seen as the absence of safety. This book, however, argues that the two conditions are unavoidably and unconditionally connected and that continued safety must encourage the continued search for safety. In other words, safe practices are found, maintained, and improved upon by actively taking risks that have the potential to increase safety. However,

organizations and individuals must be willing to take risks recognizing that the attempt may have some level of harm. This book is based on the principles that uncertainty of outcomes is always present to some degree, safety and the risk of harm are unavoidably connected, and safety choices always sacrifice one danger in favor of another. Productive safety measures come from balancing potential gains in safety with the unavoidable harm to individuals, organizations, and the environment that is brought on by those choices. The book discusses these relationships in detail and offers several strategies for increasing safety and reducing the risk of harm.

Keywords: risk, safety, uncertainty, relationships

2. How Do Disasters Evolve?

Chiles, James R. 2002. Inviting disaster: lessons from the edge of technology. New York: Harper Business. 368 p.

Annotation: While technology has provided the means for achieving unprecedented control over land, air, and sea, it has also become increasingly complex. As a result of this complexity, disasters are difficult to predict, and they are even more difficult to prevent. This book exposes many common mistakes that have culminated in unnecessary tragedies, ranging from equipment failures on oil tankers to the Challenger space shuttle tragedy. Many of the lessons learned from these tragedies have direct application to wildland firefighters, such as training for realistic emergencies, encouraging and maintaining open communication between managers and crew members, avoiding over-confidence in technology or previous experience, demanding safety despite deadlines, welcoming bad news as opportunities to catch unsafe situations, developing a healthy fear of work conditions, and focusing on small problems to avoid large problems. Managers who want to avoid technology related disasters will benefit directly from the examples and suggestions this book has to offer.

Keywords: technology, disaster, crisis, risk, training

Cohen, Eliot A.; Gooch, John. 1990. Military misfortunes: the anatomy of failure in war. New York: Free Press. 320 p.

Annotation: Military misfortunes are complex and involve not only individual failures in judgment or action, but also organizational failures rooted in the values each military organization upholds. This book uses examples from several military battles to highlight the need for military operations and organizations that: 1) *learn* from past mistakes and mistakes made by others, 2) *anticipate* future mistakes, and 3) *adapt* to changing circumstances that may reshape the type of mistakes that occur, how they occur, and when and where they may occur. The book focuses on three types of failure, moving from basic to more complex, that happen when one or more of the above three ideals are missing in an operation simple failures, aggregate failures, and catastrophic failures. This book applies directly to managers who want to better understand how failures occur and the relationship between individual and organizational responsibility of the misfortunes that result.

Keywords: military, accidents, values, organizations

Edmondson, Amy. 1996. Learning from mistakes is easier said than done: group and organizational influences on the detection and correction of human error. Journal of Applied Behavioral Science. 32(1): 5-28.

Annotation: Researchers have often studied and discussed errors and accidents within an organizational setting in two ways. The first focuses on the individual, while the second looks at the system in which the individual operates. Edmondson argues for a third perspective, one that looks at both the individual and the system, and specifically focuses on the work group. She examines the way that work groups can be used to "coordinate and to catch each other's mistakes". Edmonson examined medical errors in administering medications in two hospitals. She found that overall error incidences were lower when coordinated teams were used. This article has relevance for those interested in examining errors owing to the entire system rather than only the individual or the organization.

Keywords: leadership, groups, accidents

Maurino, Daniel E.; Reason, James; Johnston, Neil; Lee, Rob B. 1995. Beyond aviation human factors: safety in high technology systems. Brookfield, VT: Ashgate. 169 p.

Annotation: Aviation human factors investigations have typically blamed individual behavior as the primary cause of serious work accidents. However, this book argues that organizations are responsible for two aspects that contribute to work related accidents: 1) the local working conditions that restrict how workers can behave and 2) the safeguards used to protect workers from potential accidents. The authors provide a theory for making sense of disasters that predicts negative consequences that may result from organizational practices and helps investigators trace the causes of an accident back to their organizational roots. Using several case studies of aviation accidents, they show how their theory could work in practice and then clarify the steps needed to carry out this kind of analysis during an actual investigation.

Keywords: aviation, human factors, accidents

National Commission on Terrorist Attacks Upon the United States. 2004. The 9/11 commission report. Washington DC: U.S. Government Printing Office. 567 pages. Available: http://www.gpoaccess.gov/911 [July 6, 2007].

Annotation: This report is divided into 13 sections that document the incidents surrounding the 9/11 terrorist attacks. Section one focuses on homeland issues and discusses homeland defense as well as national crisis management. Section two discusses Bin Ladin and the rise of al Qaeda; more specifically, this section examines the "building, organizing, and declaring" of war on the United States by al Qaeda. Section three focuses on the evolution of counterterrorism measures throughout multiple layers of government. Sections four, five, and six discuss some of al Qaeda's initial acts of aggression as well as specific threats made to the United States. Sections seven, eight, nine, and ten examine events more directly linked to the 9/11 attacks, including the selection of terrorists, the planning of 9/11, the events of 9/11, the immediate response to the attack, and the United States plan for war. Section eleven comments on foresight and hindsight, what we should have been aware of prior to 9/11, and what we know now. Section twelve suggests a global strategy for dealing with terrorism. Finally, section thirteen offers suggestions for reorganizing America's defenses in the wake of the 9/11 tragedy. This report provides a detailed analysis of the many small organizational failures that led to the 9/11 catastrophe. In addition, the report examines the responses of emergency crews during the attacks and explores why so many first responders were killed as the World Trade Center Towers collapsed. This report has already become a classic in assessing the causes of disaster.

Keywords: 9/11, communication, strategy

Perrow, Charles. 2007. The next catastrophe: reducing our vulnerabilities to natural, industrial, and terrorist disasters. Princeton, NJ: Princeton University Press. 377 p.

Annotation: Perrow, developer of normal accident theory, argues here that we must reduce the size of targets that are vulnerable to disasters because organizations, including political ones, cannot completely prevent all the risks associated with the potential disasters that a society might face. A basic tenant of Perrow's argument is that disasters must be viewed as a normal part of our existence, for tornadoes, floods, hurricanes, earthquakes, forest fires, and even terrorists attacks, among many others, cannot be easily removed from the human environment. Perrow argues that the primary method of reducing our vulnerabilities to these inevitable disruptions is not to put a band-aide on them with simple remedies associated with prevention, remediation, and damage limitation, but to limit to the greatest degree possible, the potential damage that can be done by a devastating event. To complete this task, Perrow recommends dealing with the three major source-targets open to the most widespread devastation: 1) concentration of energy (explosive gases, toxic substances, diseased woods, and brush); 2) concentrations of populations that might be mixed with high energy concentrations; and 3) concentrations of economic and political power. Working in these three areas would have the highest payoff in reducing infrastructure damage and human casualties. The idea outlined in this book, that of reducing vulnerability to inevitable disasters, may be helpful to the managers of high risk organizations who are responsible for preventing and responding to disasters.

Keywords: disaster prevention, politics, normal accident theory

Perrow, Charles. 1999. Normal accidents: living with high-risk technologies. Princeton, NJ: Princeton University Press. 386 p.

Annotation: Many of the accidents that organizations face are a result of complex interactions between multiple events and with multiple actors. They cannot be explained as being only one group or individual's "fault". In this book, Perrow investigates the complexity of accidents as events that are inevitable because of the complex interactions between actors, organizations, procedures, supplies, environment, and equipment. Often these elements combine in ways that actually make the problem bigger than it would have been if they had occurred alone. For example, a broken piece of equipment may not cause too much trouble on its own, but if that element were combined with a high-stress situation that demanded the use of that equipment by an untrained worker, the situation could very quickly get out of hand. Perrow explores the complexities and patterns of such accidents using examples from several high risk industries including: nuclear power, chemical, airline, marine transport of dangerous materials/weapons, space, and gene splicing. Recognizing the complexity of routine "accidents" is the first step to untangling the contributing elements. Without recognizing what contributes to these situations, it is difficult to prevent them from happening again. Readers are reminded that managers faced with the task of preventing and responding to disasters must avoid over-simplifying them by looking at the "system," and by recognizing the complex interdependent contributions of key events and players in the culmination of a disaster.

Keywords: accidents, accident prevention, risk, disaster, normal accident theory

Snook, Scott A. 2000. Friendly fire: The accidental shootdown of U.S. Black Hawks over Northern Iraq. Princeton, NJ: Princeton University Press. 257 p.

Annotation: On April 14, 1994, two United States (U.S.) Air Force fighters accidentally shot down two U.S. Army Black Hawk helicopters over Northern Iraq. Military and civilian investigations into the tragedy soon followed. After nearly 2 years, neither investigation had elicited a definitive cause or culprit. In this book, Snook seeks to explain the causal chain of events that led to this disaster. The book is organized into seven chapters. Chapters one and two provide an introduction to the book as well as an explanation of what happened. Chapters three, four, and five explain different pieces of the puzzle at individual, group, and organizational levels of analysis. Chapter six takes a holistic

approach to the disaster and discusses Snook's concept of "practical drift," which is the slow uncoupling of practice from written procedure. Chapter seven summarizes the lessons learned from the various levels of analysis. Snook concludes by offering both practical and theoretical implications for understanding the causes of tragedy.

Keywords: accidents, military

Starbuck, W. H.; Milliken, F. J. 1988. Challenger: fine-tuning the odds until something breaks. Journal of Management Studies. (25): 319-340.

Annotation: The authors of this article discuss the tragic Challenger disaster of 1986. They claim that investigations of the Challenger accident show the ways in which organizations can be unrealistic and error-prone, including errors in judgment, communication, control, and generalizations about past successes. The authors use NASA and Thiokol as organizations that illustrate the effects that generalizations about past successes can have on organizations. Three theories are offered regarding future successes and the effects of experience: 1) neither success nor failure changes the expected probability of a subsequent success, 2) success makes a subsequent success seem less probable, and failure makes a subsequent success appear more likely, and 3) success makes a subsequent success appear more probable, and failure makes a subsequent success seem less likely. In conclusion, the authors offer the idea that disasters can actually help organizations because they can learn how to reduce the cost of failures, prevent the repetition of failures, and make failures rarer. Learning from disasters, however, involves looking beyond initial reactions and explanations and addressing less likely causes.

Keywords: Challenger, disaster, judgment

Tompkins, Phillip K. 2005. Apollo, Challenger, Columbia: the decline of the space program. Los Angeles: Roxbury. 288 p.

Annotation: Tompkins takes a case study approach to one organization, the National Aeronautics and Space Administration (NASA), while simultaneously highlighting topics such as culture, identification, leadership, and structure that are relevant to all organizations. Tompkins integrates communication theory and concepts in his discussion of the history of the space program and the lessons the decline of this program has for all organizations. More specifically, this book identifies 11 "communication transgressions" common to organizations and individuals in trouble. Tompkins concludes by contrasting the practices of the space program with practices and ideologies of two healthy organizations, and discusses the importance of ethics in today's environment of corporate distrust by the public.

Keywords: communication, NASA, Challenger, Columbia

Turner, Barry A.; Pidgeon, Nick F. 1997. Man-made disasters. 2nd ed. Boston: Butterworth-Heinemann. 250 p.

Annotation: To understand and avoid future calamities, decision makers must have a more accurate way of understanding past calamities. Most of what we know about calamities comes from eye witness accounts that favor relief efforts and damage reports rather than the specific events that come together initially to form a disaster. This suggests that previous models of disaster may be inaccurate. This book looks for similarities between numerous disasters and examines social-, psychological-, and communication-based research to form a more accurate model of the events that create and propel disasters. The authors offer suggestions to guide organizations and managers in developing a culture of safety in an environment that promotes learning from past mistakes.

Keywords: disaster, safety culture

Turner, Barry. 1994. Causes of disaster: sloppy management. British Journal of Management. 5: 215-219.

Annotation: Turner argues that while the best way to avoid disasters is primarily "for managers to establish, to strengthen, and then to assert control," management control only addresses part of the problem, and there are limitations that affect management in disaster situations. Turner advocates examining three interrelated factors in a disaster situation: the technical aspect of the disaster, the administrative aspect, and managerial issues. Turner notes some of the ways in which administrators can see clues that management is becoming sloppy in ways that might increase the chances of a disaster. He concludes that there is no way to avoid all disasters; however, being aware of specific strategies can help minimize the costs and damages of disasters. Some of these strategies include: avoiding rigid orthodoxy; improving communication with and between staff; understanding reasons for outside complaints; and acknowledging potential consequences or company actions.

Keywords: management, accidents, control, disaster prevention

Vaughan, Diane. 1996. The Challenger launch decision: risky technology, culture, and deviance at NASA. Chicago, IL: The University of Chicago Press. 592 p.

Annotation: See annotation in Section II.A.1. *How do people make decisions?* p. 19.

Weick, Karl. 1991. The vulnerable system: an analysis of the Tenerife air disaster. In: Frost, Peter J.; Moore, Larry F.; Louis, Meryl R.; Lunberg, Craig C.; Martin, Joanne, eds. Reframing organizational culture. Newberry Park, CA: Sage: 117-130.

Annotation: See annotation in Section III.B.2. *What can we learn from high reliability organizations?* p. 61.

3. How Do Managers Deal With Risk/Uncertainty?

Alder, G. Stoney. 1997. Managing environmental uncertainty with legitimate authority: a comparative analysis of the Mann Gulch and Storm King Mountain Fires. Journal of Applied Communication Research. 25: 98-114.

Annotation: Alder recognizes two decisions common to both the Mann Gulch and Storm King Mountain fires that influenced the behavior of firefighters during critical moments: 1) failing to question authority and 2) failing to obey authority. He argues that these failures are based on individual perceptions of legitimate authority. There are four issues contributing to decreasing perceptions of legitimate authority between crew members and their leaders, including lack of prior experience, positive mutual interaction, communication of procedural explanations, and communication of process control. Managers who want to increase the legitimacy of their authority in the eyes of their team members should learn from the mistakes made by leaders and participants in both these fires.

Keywords: authority, Mann Gulch, Storm King, communication

Argyris, Chris. 1990. Overcoming organizational defenses: facilitating organizational learning. Needham Heights, MA: Allyn and Bacon. 180 p.

Annotation: See annotation in Section II.E.1. *What is organizational learning?* p. 39.

Carroll, John S. 1998. Organizational learning activities in high-hazard industries: the logics underlying self-analysis. Journal of Management Studies. 35: 699-717.

Annotation: See annotation in Section II.E.1. *What is organizational learning?* p. 39.

Cleaves, D. A.; Haynes, R. W. 1999. Risk management for ecological Stewardship. In: Sexton, W.T.; Malk, A.J.; Szaro, R.C.; Johnson, N.C., eds. Ecological stewardship: a common reference for ecosystem management. Vol. 3. Oxford, UK: Elsevier Science, Ltd: 431-461.

Annotation: This comprehensive chapter documents, from a management perspective, the knowledge base on risk assessments and risk management. The previous chapter in the book is a companion article that provides the scientific foundation for the concepts and terminology used by Cleaves and Haynes. In this chapter, Cleaves and Haynes provide a framework for considering the uncertain outcomes of management actions and natural events. They review uncertainty and risk management principles, review strategies for adjusting ecosystem risks, and provide suggestions for fitting uncertainty management into organizational cultures. They begin with a definition of terms. For example, *uncertainty* is described as a set of unknowns, based on a combination of ambiguity, knowledge gaps, and variation in natural and human processes. *Risk*, which is the likelihood of an adverse effect or loss, reflects the human assessment of value. *Risk assessment* is used to estimate the probabilities and magnitudes of the effects and consequences (gains and losses) of stressors on natural resources. *Risk adjustment* is an effort to modify, mitigate, or respond to stressors in a way that holds consequences to acceptable levels. *Risk characterization* is a summary of the risk, its nature and context, and how it might be adjusted. *Risk communication* is an exchange of information about a particular risk or group of risks. Most of the article describes the risk management cycle, including detailed discussions of hazard identification, risk assessment, evaluation, adjustment, implementation, and monitoring. The authors use case studies to illustrate issues of endpoint objectives, expert judgment, cost/benefit trade-offs, risk adjustment instruments (for example, performance standards, regulations, incentives), and shifting risk profiles. They also describe an example of a USDA Forest Service Decision Protocol designed for use by Interdisciplinary Teams. The final sections address risk communication, the adoption of a risk management paradigm, and risk-based thinking in organizations. The authors recognize that, rather than documenting decisions already made, good risk assessments expose many incorrect and unfounded assumptions, stimulate learning, and improve decision making. They emphasize embracing uncertainty, accepting risk, using failures to stimulate learning, using probabilities for risk assessment, viewing risk management as iterative and adaptive, ensuring that risk assessments are actually applied to decision making, and developing organizational characteristics that promote risk-based thinking.

Keywords: risk management, risk assessment, risk adjustment, uncertainty, risk communication

Columbia Accident Review Board. 2003. Columbia Accident Review Board Report. Vols. I-VI. Washington, DC: Government Printing Office. 341 p. Available: http://caib.nasa.gov [July 6, 2007].

Annotation: See annotation in Section II.A.3. *Sensemaking and crisis.* p. 20.

Ewing, Lance J.; Lee, Ryan B. 2004. Surviving the age of risk: a call for ethical management. Risk Management Magazine. 51(9): 56-58.

Annotation: Ewing and Lee look at some of the ways to consider ethical risk management in a corporate context, which have changed because of recent scandals such as Enron. They give six ways to create an ethical risk management environment (the six Cs): 1) *Champions*: find a spokesperson, someone to "champion" risk management; 2) *Commitment*: make a commitment for this to happen and actually take hold within the organization; 3) *Consistency*: maintain consistency throughout departments in the organization; 4) *Correlations*: pay attention to the ways in which a variety of issues impact risk management; 5) *Communication*: communicate between departments and with each other about

the risks that are possible; and 6) *Code of Ethics*: have a code of ethics that can help the organization see some of the common (or not so common) concerns that members of the organization might have. The authors conclude by stating that there can only be ethical risk management if there is a desire to shift the ways in which the organization thinks about risk management, and that there must be a change in the way members of risk management communities view themselves as an organization and in the larger corporate community.

Keywords: risk management, risk, ethics

Hopkins, Andrew. 1999. Managing major hazards: the lessons of the Moura Mine disaster. St. Leonards: Allen & Unwin. 172 p.

Annotation: In every organization, things go wrong. For the most part, these errors are minor and often go unnoticed. However, when disaster occurs, external pressure often forces the exposure of many of the failures that occur within an organization. Thus, a disaster can offer an opportunity for in-depth analysis of the internal workings of an organization. The Moura mine explosion is one such disaster. In this book, Hopkins argues that the explosion was the result of a series of system, cultural, hierarchical, and managerial errors. Furthermore, he argues that the organization was "systematically inattentive to the potential for errors". In order to correct this organizational shortcoming, steps must be taken to address the aforementioned series of errors. This book begins with a brief preview of literature on disasters that highlights organizational factors pertinent to the Moura mine disaster. Next, the roles of communication failure and organizational culture are discussed. Subsequent chapters cover management shortcomings, specifically targeting the issue of safety through safety audits, measures of safety, an emphasis on productivity over safety, and management risk assessment. Additionally, the issue of safety is discussed on a broader level, and the common argument "safety pays" is discussed. The book concludes with a discussion of regulatory systems that are most appropriate for the coal mining system, and Hopkins summarizes the organizational failings and lessons revealed by the Moura mine explosion.

Keywords: disaster, safety, communication, culture

Hopkins, Andrew. 2000. Lessons from Longford: the Esso Gas Plant explosion. Sydney: CCH Australia Limited. 184 p.

Annotation: The Esso gas plant explosion of 1998 represents a series of organizational failures that resulted in devastating consequences, including two deaths. This book examines those organizational failures through the findings of the Royal Commission. The author argues that the accident was preventable and was caused by a number of failures including the failure to heed warning signs, communication problems, insufficient attention to hazards, deficient auditing, and a failure to learn from past experiences. Due in part to these failures, the Royal Commission found that fault for the explosion lay not with the plant operators but with Esso. The failures that led to the explosion are not atypical organizational failures, and thus an analysis of the factors leading to this explosion bears implications for other organizations facing major hazards. A number of themes are integrated into this book, including practical preventability, accident cause, operator error, market forces, governmental influence, organizational analysis, high reliability organizations, and a comparison to the Moura mine disaster.

Keywords: high reliability organizing, accidents

Kahneman, Daniel; Slovic, Paul; Tversky, Amos, eds. 1986. Judgment under uncertainty: heuristics and biases. New York: Cambridge University Press. 544 p.

Annotation: This is a classic textbook written by three well known authors (Kahneman recently won the Nobel Prize for economics) who have spent their careers working in the psychological fields of understanding how people make decisions under uncertainty. The introduction to this edited volume identifies three heuristics (in other words, rules of thumb) that are employed in making judgments under uncertainty: *representativeness*, *availability*, and *adjustment* from an anchor. *Representativeness* is used when people are asked to judge whether one thing belongs to another thing, is representative of another thing, or resembles another thing. *Availability* is used when people assess the frequency or probability of an event by the ease in which occurrences come to mind. *Adjustment* from an anchor is used when people make numerical estimates by starting from an initial available value. While these heuristics can be effective, the authors argue that they can also lead to predictable biases and errors. Therefore, they propose a better understanding of these heuristics and their biases in order to improve judgments and decisions in uncertain situations. There are 10 parts in this volume, which aim to provide a comprehensive sample of judgmental heuristics and their effects. The first part provides a review of heuristics and biases of intuitive judgments. Part II discusses the representative heuristic and Part III deals primarily with problems of causal attribution (extending from the representative heuristic). Part IV discusses the availability heuristic and social judgment. Part V illustrates the perception and learning of judgments of lay people and experts. Part VI describes probability assessors and overconfidence in prediction and explanation. Part VII discusses multistage inference biases. Part VIII describes procedures for correcting and improving judgment. Part IX summarizes the effects of judgmental biases in the perception of risk. And finally, Part X discusses conceptual and methodological issues involved in the study of heuristics and biases. This book is oriented toward academic audiences, but that shouldn't sway people from looking into it, particularly if they want to gain a deeper understanding of the heuristics and biases. Notable chapters include those on learning from experience, hindsight bias, and understanding perceived risk.

Keywords: uncertainty, risk, judgment, heuristics, bias

Krieger, Janice L. 2005. Shared mindfulness in cockpit crisis situations. Journal of Business Communication. 42(2): 135-167.

Annotation: Research reveals that human error contributes 60 to 80 percent of error in aviation accidents and disasters. Thus, despite innovations in technology and safety materials, individuals must be able to make speedy yet intelligent decisions and be able to communicate those decisions in an efficient manner. Krieger explores the psychological construct of shared mindfulness, specifically how it is constructed through communication and used. The study's findings support the existence of shared mindfulness as a construct, reveal process categories that create shared mindfulness, and show that dyads that employed more behaviors indicative of shared mindfulness made more effective decisions.

Keywords: shared mindfulness, crisis communication, decision making, mindfulness

Lagadec, Patrick. 1993. Preventing chaos in a crisis: strategies for prevention, control and damage limitation. Berkshire, England: McGraw Hill Book Company. 363 p.

Annotation: This early work of Lagadec's, though it is nearly a decade and half old, continues to provide a helpful summary of guidelines mangers can use during a crisis. The book, targeted specifically for decision makers, is organized into three parts that answer three main questions: What is a crisis? How do you manage a crisis? How do you develop a learning process to prevent a crisis? Case studies from the Exxon-Valdez oil spill, the radioactive threats posed by Three Mile Island and Chernobyl nuclear accidents, the 1985 Mexico earthquake, the Tylenol pill tampering case, and the Cuban missile crisis, among many others, are used to drive home his crucial points. Lagadec is describing how to have a "brutal audit" to anticipate potential errors before the real-time brutal accident occurs, and how to properly prepare a course of action, both individually and organizationally, once chaos and disaster do occur.

Keywords: crisis management, crisis prevention

Leveson, Nancy. 2004. A new accident model for engineering safer systems. Safety Science. 42: 237-270.

Annotation: Leveson argues that most accident models are designed for simple systems. Newer accident models are needed because of the changing landscape of organizational systems and the changing contexts in which they are developed. Fast-paced technological change, new types of digital failure and hazards, decreasing tolerance for failure, and new regulatory standards all necessitate a new paradigm of accident model. Leveson meticulously examines older event-based models and their limitations. She argues that event-based models are often subjective and open to bias. They are poor at recognizing management or cultural factors and often do not account for personal adaptation and human error. Leveson argues for a new model based on systems theory. In this new model, safety is viewed as a control problem and is managed by a control system that is adaptive and socio-technical. Additionally, she discusses how the new model will impact accident analysis, accident prevention, risk assessment, and performance monitoring.

Keywords: accident models, systems theory, accident prevention, risk assessment, performance

Reason, James. 1997. Managing the risks of organizational accidents. Brookfield, VT: Ashgate. 252 p.

Annotation: This book focuses on the causes, consequences, and possible means of avoiding organizational accidents. While individual accidents are more frequent and often target the individual for blame, organizational accidents are deep rooted errors in the daily function of organizations that increase the likelihood for disaster. The author discusses the relationship between organizational hazards caused by technology and human factors and possible methods that organizations can use to defend against those hazards. He uses disasters in numerous industries to illustrate how organizational accidents occur and how to prevent them. Lastly, he argues that organizations must provide and cultivate a culture of safety if they are to decrease the possibility of on-site accidents, he provides three models for doing this, and he discusses the consequences for workers and organizations of neglecting safety.

Keywords: accidents, safety culture, disaster, accident prevention

Slovic, Paul; Finucane, Melissa; Peters, Ellen; MacGregor, Donald G. 2002. Risk as analysis and risk as feelings: some thoughts about affect, reason, risk, and rationality. Risk Analysis. 24(2): 1-12.

Annotation: Risk assessment is often viewed as a logical, cause-effect process that uses mathematics and gives little credence to feelings. Without discounting the need for a rational approach to categorizing risk analysis, the authors show how affect, feelings, and emotions (the perceived "goodness" and the "badness" of a risky situation) can impact how individuals perceive and evaluate risk, and ultimately, how they decide on a particular course of action. The authors argue that it is not only what people think about an activity or technology, but how they feel about it that is also important. They call these feelings about risk the *affect heuristic.* Each person holds an *affects pool* in his/her head that contains all of his/her positive and negative markers, and the individual consults this affects pool when making a decision. The authors support their argument in viewing the affect heuristic as a necessary component of risk analysis with a thorough reading of the literature on risk and feelings and by describing numerous empirical studies and experiments that show how the affect heuristic led to certain outcomes. Some examples described in the paper of how the affect heuristic affects decision making are drawn from finance (picking "good" stocks to invest in from "bad" ones); quitting smoking (how the enjoyment of smoking can over-ride its health risks); and toxicology (where toxicologist's judgments about potentially dangerous chemical risks was show to be influenced by affective processes). The authors describe the difference between a *possibility* and a mathematical *probability*, how reliance on affect can be misleading, and how affect can be manipulated through advertisements and marketing. The paper's conclusion is blunt: "We cannot assume that an intelligent person can understand the meaning of and properly act upon even the simplest numbers…or statistics pertaining to risk, unless these numbers are infused with affect". This paper, though written for an academic audience, is very readable, and has many potential applications for understanding how feelings mesh with the risk assessments managers now use in their daily management operations.

Keywords: risk assessment, heuristics

Staw, Barry; Sandelands, Lance; Dutton, Jane. 1981. Threat rigidity effects in organizational behavior: a multilevel analysis. Administrative Science Quarterly. 26(4): 501-524.

Annotation: This article addresses how organizations deal with adversity and how organizations adapt within adverse conditions. The authors comment that most research emphasizes "organizational and not individual or group responses to adversity," and those studies tend to "take a functional stance". What those approaches leave out is the potential for "maladaptive or pathological cycles of behavior" that may obstruct productivity within the organization. Particularly within situations that are threatening to the organization or people within it, there may be a higher sense of rigidity, using previous methods of adaptation, which may not be the most beneficial. The authors look at ways in which individuals, groups, and organizations are affected during threat situations. By examining these effects from a number of levels, managers are able to see how threat affects each level and potentially address issues at all levels simultaneously, rather than potentially having a system breakdown among the various levels that are affected by threat-situations.

Keywords: organizational behavior, organizations

Tompkins, Phillip K. 1990. On risk communication as interorganizational control: the case of the aviation safety reporting system. In: Kirby, Andrew, ed. Nothing to fear: risks and hazards in American society. Tucson, AZ: University of Arizona Press: 203-239.

Annotation: Focusing on the 1974 Trans World Airlines (TWA) Flight 514 crash, Tompkins discusses the National Transportation Safety Board's (NTSB) investigation of the crash and points to communication between the pilot and controllers and communication between airline organizations as principle contributors to the tragedy. Using this analysis, he critiques Perrow's book "Normal Accidents: Living with High-Risk Technologies" (see annotation of 1999 reprint in this section) and highlights how current hazard research has ignored the central role communication plays in disasters. Promoting policies and behaviors that reinforce a culture of open communication may help prevent tragedies like the TWA Flight 514 crash.

Keywords: risk communication, disaster, TWA Flight 514

Tompkins, Phillip K. 1993. Organizational communication imperatives: lessons of the space program. Los Angeles: Roxbury. 238 p.

Annotation: Drawing upon experience working for NASA during the Apollo Missions and his studies of organizational communication, Tompkins illustrates that taking a communication perspective can help with understanding organizational problems. This book takes a narrative approach in which Tompkins discusses his visits to the Marshall Space Flight Center as a consultant, as well as his assessment of the Challenger accident as it was described by the Roger's Commission Report. Tompkins explains communication practices, such as upward communication, interorganizational penetration, and "Monday Notes," that made the Marshall Flight Space Center so successful during NASA's Apollo Missions. Tompkins then contrasts the best practices of NASA during the Apollo era with practices that led to the Challenger tragedy. Tompkins' book reveals organizational communication practices that lead to success or failure in organizations, such as NASA, that have little room for error.

Keywords: NASA, leadership, communication, Challenger

Weick, Karl; Sutcliffe, Kathleen. 2007. Managing the unexpected: resilient performance in an age of uncertainty. 2nd ed. San Francisco, CA: Jossey-Bass. 224 p.

Annotation: See annotation in Section III.B.2. *What can we learn from high reliability organizations?* p. 60.

B. High Reliability Organizing

While some organizations and employees never have to deal with crisis or disaster, other organizations, such as high reliability organizations (HROs), are preoccupied with them. Even though these organizations operate in a constant state of heightened awareness, they still reliably avert problems. For those interested in applying these concepts to wildland fire organizations, this section introduces the HRO and some of its defining characteristics (Bierly and Spender 1995; LaPorte 1996). Several readings directly address the concept of mindfulness (Knotek and Watson 2006; Langer 1997; Weick and Putnam 2006; Weick and Sutcliffe 2007; White, in press). Case studies of HROs add depth on what makes these organizations unique and they demonstrate lessons learned from HROs (White, in press). Lessons include general management techniques (Keller 2004; Weick and Sutcliffe 2007), as well as more specific recommendations for creating culture (Klein and others 1995), encouraging employee assimilation (Meyers 2005), and making decisions (Roberts and others 1994). Several

readings also address the differences between normal accident theory and high reliability theory (Rijpma 2003; Laport and Rochlin 1994); in light of this debate, Rijpma (2003) summarizes several other readings included in this annotated reading list.

Author's Picks

- Keller, Paul, technical writer-editor. 2004. Managing the unexpected in prescribed fire and fire use operations: a workshop on the high reliability organization. Santa Fe, New Mexico, May 10-13, 2004. Gen. Tech. Rep. RMRS-GTR-137. Fort Collins, CO: U.S. Department of Agriculture, Forest Service, Rocky Mountain Research Station. 73 p. Available: http://www.wildfirelessons.net/HRO.aspx [July 6, 2007].

- Roberts, Karlene. H., ed. 1993. New challenges to understanding organizations. New York: Macmillan. 256 p.

- Roberts, Karlene H. 1990b. Managing high reliability organizations. California Management Review. 32(4): 101-113.

- Weick, Karl; Sutcliffe, Kathleen. 2007. Managing the unexpected: resilient performance in an age of uncertainty. 2nd ed. San Francisco, CA: Jossey-Bass. 224 p.

1. What is High Reliability Organizing?

Bierly, Paul E; Spender, J.-C. 1995. Culture and high reliability organizations: the case of the nuclear submarine. Journal of Management. 21(4): 639-656.

Annotation: Organizational theorists present contrasting ideas regarding how to define a *high risk organization* and the *normality of accidents* within such organizations. This paper examines one type of high risk organization, a United States (U.S.) Navy nuclear submarine's program, and its culture. Bierly and Spender found that the strict learning-and-training culture of the U.S. Navy, combined with the stringent command-and-control structure of submarine operations, has produced high reliability for this organization. The nuclear submarine illustrates how culture and experience can interact with bureaucracy to turn a potentially high risk system into a high reliability system. When a submarine is in action, there is a strict hierarchy and a specific delegation of responsibilities. This delegation of responsibilities is controlled and enforced via selection, training and mutual monitoring, criticism, and advice. This combination of factors yields an efficient communication system with the ability to react to, and absorb, damages and surprises. Through such practices, the U.S. Navy nuclear submarine program maintains a high level of reliability.

Keywords: military, culture, reliability, training

LaPorte, Todd. 1996. High reliability organizations: unlikely, demanding, and at risk. Journal of Contingencies and Crisis Management. 4(2): 60-71.

Annotation: Although many scholars argue that accidents are inevitable in organizations, studies of high reliability organizations (HROs) challenge this characterization. These organizations deal with complex technologies on a daily basis without major crises and are still able to meet high demands. Through the HRO project, LaPorte investigates the central characteristics of high reliability organizations. He gives a brief overview of the provisional findings of the HRO project. First, he discusses the internal processes of HROs, including organizationally defined intent, reliability enhancing operations, and an organizational culture of reliability. Next, he discusses external relationships and argues that independent public bodies and stake-holding interest groups are crucial to HRO success. These external stakeholders create boundary-spanning processes through which reliability is encouraged. Finally, LaPorte discusses the future challenges of HROs as they increase in number and importance.

Keywords: high reliability organization, stakeholders, crisis

Rijpma, Jos A. 2003. Book review essay. From deadlock to dead end: the normal accidents-high reliability debate revisited. Journal of Contingencies and Crisis Management. 11(1): 37-45.

Annotation: Rijma reexamines the debate over the inevitability of accidents in modern, high risk systems through the paradigms of *Normal Accident Theory (NAT)* and *High Reliability Theory (HRT)*. *NAT* proposes that serious accidents and disasters are inevitable in modern, high risk technological systems. Conversely, *HRT* points out that high reliability organizations have achieved commendable safety records despite the hazards inherent to the technological systems they utilize. In recent years, both NAT and HRT have further evolved into encompassing perspectives that provide frameworks for understanding and preventing accidents. Rijpma summarizes a vareity of books and articles (including several in this annotated reading list) that have attempted to direct this debate in one direction or the other in their assessments of various disasters. This review essay ties these developments together to depict the current state of the debate. Despite recent literature and the evolution of NAT and HRT, Rijma argues that the "debate is not only at a deadlock, but has actually reached a dead end in itself. Any progress within the two perspectives is possible, but progress along the lines of this debate is further away than ever". Rijpma concludes that the debate is political and will not further the scientific understanding of accidents.

Keywords: normal accident theory, high reliability theory, accidents, disaster, accident prevention

Roberts, Karlene. H., ed. 1993. New challenges to understanding organizations. New York: Macmillan. 256 p.

Annotation: This book is designed to raise manager awareness of factors that could prevent failures within their organizations. The book begins by defining and discussing common characteristics of high reliability organizations. Next, it presents a multi-dimensional model of reliability-seeking organizations, where a decomposable system is contrasted with a holistic system with regard to reliability. Next, the book places reliability as central to effectiveness research and addresses the relationship between safety and reliability. The final chapters address issues relevant to understanding organizations, including the impact of ideology on organizational behavior and the difficulties associated with changing ideology, especially in a dynamic environment. Additionally, the book discusses the impact of environmental characteristics on organizational uncertainty in situations requiring fast decision making. It also addresses the impact of fatigue on crew performance and the necessity of communication to creating a safe, reliable organization. The book concludes with a chapter presenting an applied research program to reduce accidents in an organization where disasters come at a high price, and a case study developed as a basis for the underlying tool (probabilistic risk assessment) to the research program. The chapters incorporate real-life examples and narratives to illustrate points ranging from President Reagan's firing of the air traffic controllers to the Tenerife air disaster. It addresses multiple issues associated with organizational failures to facilitate a more complete understanding of the complexity of organizations and the factors that lead to disaster.

Keywords: reliability, organizational behavior, communication, fatigue, risk assessment, Tenerife air disaster, accident prevention, performance, uncertainty

Roberts, Karlene H. 1990a. Some characteristics of one type of high reliability organization. Organization Science. 1(2): 160-176.

Annotation: Roberts argues that while the primary goal of many organizations is not performance reliability, in high reliability organizations (HROs), such as nuclear powered aircraft carriers, performance reliability is crucial because of the possibility of catastrophic error. She makes the case that current accident and crisis literature does not fully account for the distinctiveness of HROs. Previous crisis research by Perrow and Shrivastava argues that if dysfunctional components of high risk technology exist, catastrophe is bound to occur. To respond to this, over the course of 3 years, a research team engaged in participant observation on board a nuclear powered aircraft carrier. From the resulting data, Roberts argues that dysfunctional components do exist on aircraft carriers, but are skillfully handled. For example, while Perrow argues that complexity and tight coupling in organizations is extremely risky, Roberts argues that aircraft carriers are able to deal with these issues regularly with reduced risk. Similarly, while Shrivastava notes that issues like poor training are antecedents to accidents, Roberts argues that aircraft carriers have strategies in place in order to coexist with such problems. The article concludes with suggestions for future research on HROs.

Keywords: high reliability organization, reliability, performance

Roberts, Karlene H. 1990b. Managing high reliability organizations. California Management Review. 32(4): 101-113.

Annotation: In this introductory article, Roberts investigates how high reliability organizations are managed. She uses three specific case examples: the Pacific Gas and Electric Company, the Federal Aviation Administration's Air Traffic Control Centers, and U.S. Naval aircraft carriers. She provides a brief background on these three organizations and then details how they manage complexity, such as complex technology, systems that serve incompatible functions, indirect information sources, and baffling interactions. She then discusses how the three organizations manage tightly coupled, or mechanistic, technologies. This is achieved by the use of time dependent processes, invariant sequencing of operations, multiple ways to reach a goal, and small amounts of slack. Roberts also discusses how such organizations reduce overall risk and the high cost of risk reduction strategies.

Keywords: high reliability organization, managing complexity, aviation

Rochlin, Gene. I. 1993. Defining high reliability organizations in practice: a taxonomic prologue. In: Roberts, Karlene H., ed. New challenges to understanding organizations. New York: Macmillan: 11-32.

Annotation: The purpose of this essay is to "identify and characterize both the static and dynamic characteristics of organizations" that have not just avoided disasters, but actively reduced their probability of failure in an environment replete with error. Rochlin recants 11 characteristics that either solely, or in conjunction with one another, are indicative of a high reliability organization. These characteristics are: 1) the assumption that errors are omnipresent and thus a successful organization is a vigilant organization; 2) an assumption that sources of error are dynamic, and thus in addition to vigilance, the organization must be continually reenergized; 3) as a result of characteristics 1 and 2, an understanding that the operating environment is a constant threat and requires attention during times of disaster and when things are going well; 4) maintenance of modes of problem solving as well as avoiding the pitfall of adopting a "best" approach to resolve a problem; 5) structural flexibility adapted to the nature of the contingencies; 6) an organizational commitment to developing modes to both predict and react to current and potential problems; 7) the empowerment of organizational units dedicated

to discovering error and the potential for error; 8) an unwillingness to test reliability boundaries, or an emphasis on trial-and-error as secondary as opposed to the primary learning mode; 9) the inexistence of "stopping rules" for self-improvement and self-regulation as long as organizational resources and time provide them; 10) leniency with formal codes and rules extended with accepted standard operating procedures; and 11) agreement with the proposition that even if a complete formal history and analysis were available, the search for error would only be simplified, not removed or reduced in importance. In sum, these organizations "seek an ideal of perfection but never expect to achieve it". Importantly, Rochlin notes that the organizations he studied implemented their own organizational cultures of safety and reliability and thus belie traditional advice or models on organizational effectiveness.

Keywords: organizational effectiveness, high reliability organizations

Schulman, Paul. R. 1993a. The analysis of high reliability organizations: a comparative framework. In: Roberts, Karlene H., ed. New challenges to understanding organizations. New York: Macmillan: 33-54.

Annotation: Much research to date has focused on discovering common criteria for maintaining high reliability while managing complex, hazardous technologies. To further understand organizational management, Schulman discusses differentiation within a set of high reliability organizations. Schulman focuses on how the technical character of hazard influences both the management of risk and the maintenance of high reliability. He offers a comparative framework for examining differing approaches to achieving high reliability. One critical difference is the state and behavior of the system in operation versus in failure. Labeling organizations based on safety requirements under failure allows for a classification system for high reliability organizations. This classification system has four basic dimensions: 1) the decomposable organization, characterized by localized levels of action and analysis required for safety; 2) the holistic organization, characterized by system-wide levels of action or analysis required for safety; 3) the action-focused organization, characterized by localized levels of action and a system-wide level of analysis required for safety; and 4) the clearance-focused organization, characterized by both system-wide levels of action and analysis required for safety.

Keywords: high reliability organizations, systems thinking

Schulman, Paul. 1993b. The negotiated order of organizational reliability. Administration and Society. 25(3): 353-372.

Annotation: Schulman looks at the term *slack* as a "critical, if underappreciated, managerial resource". He defines *slack* as extra time or resources that can be used by the organization. He argues that slack has decreased in organizations and therefore, organizations have a smaller margin of error. Using Diablo Canyon as a case study, Schulman argues that there is a third type of slack, conceptual slack, where organizational members diverge in their thoughts on how the organization should be handling a situation. This type of slack can increase accountability of the organization to its members. He argues that an organization's reliability may also increase, given the potential for a broader range of choices that are available to that organization. In a time where the other two forms of slack are decreasing, he argues that increasing conceptual slack can help an organization's ability to maintain safe standards.

Keywords: performance, rules, Diablo Canyon, choice

Weick, Karl. 1987. Organizational culture as a source of high reliability. California Management Review. 29(2): 112-127.

Annotation: See annotation in Section II.B.2. *How do cultures impact organizations?* p. 24.

White, Cindy, ed. In Press. Special issue on high reliability organizations. Fire Management Today. 68(1).

Annotation: This special issue of "Fire Management Today" was compiled to illustrate how the wildland fire community is currently using high reliability organizing principles. Article topics include building a foundation for a learning culture, making sense of high reliability and learning, the genesis and evolution of high reliability organizing, lessons learned from firefighters, the first Managing the Unexpected Workshop, several personal accounts and case studies written by members of the fire community, spreading the word on high reliability, teaching about mindfulness, and an effort to assess high reliability in the wildland fire community.

Keywords: high reliability organizing, learning, mindfulness

2. What Can We Learn From High Reliability Organizations?

Eisenhardt, Kathleen M. 1993. High reliability organizations meet high velocity environments: common dilemmas in nuclear power plants, aircraft carriers, and microcomputer firms. In: Roberts, Karlene H., ed. New challenges to understanding organizations. New York: Macmillan: 117-136.

Annotation: High reliability organizations have a low tolerance for mistakes because a single mistake can create overwhelming negative consequences to human life and organizational property. However, danger is not the only thing that can impact organizations. This author argues that organizations must account for the environment in which they exist when managing risk. *High velocity environments* are characterized by rapid changes in organizational conditions such as market demand, competitors, technology, and/or regulations that quickly make information inaccurate, unavailable, or obsolete. Since information is needed to make

productive decisions, such environments place unique pressures on both organizations and their members. This article distinguishes high reliability organizations from high velocity environments and points out three similarities between the two: 1) the need for task specialization; 2) the use of rich real-time information; and 3) the balance between centralizing decision making power in upper-level management while at the same time, giving ground level workers the power and responsibility to make important safety related decisions.

Keywords: high reliability organization, high velocity environment, risk, safety

Hopkins, Andrew. 2000. Lessons from Longford: the Esso gas plant explosion. Sydney: CCH Australia Limited. 184 p.

Annotation: See annotation in Section III.A.3. *How do managers deal with risk/uncertainty?* p. 52.

Keller, Paul, technical writer-editor. 2004. Managing the unexpected in prescribed fire and fire use operations: a workshop on the high reliability organization. Santa Fe, New Mexico, May 10-13, 2004. Gen. Tech. Rep. RMRS-GTR-137. Fort Collins, CO: U.S. Department of Agriculture, Forest Service, Rocky Mountain Research Station. 73 p. Available: http://www.wildfirelessons.net/HRO.aspx [July 6, 2007].

Annotation: See annotation in Section I. *Human factors and firefighting.* p. 8.

Klein, Rochelle Lee; Bigley, Gregory A.; Roberts, Karlene H. 1995. Organizational culture in high reliability organizations: an extension. Human Relations. 48(7): 771-793.

Annotation: This article sheds light on cultural similarities and differences between high reliability organizations and other organizations, and helps differentiate between high reliability organizations. The authors take a widespread comparative approach of culture assessments in multiple organizations. They compare the culture assessments in two high reliability organizations with like assessments in high reliability organizations and various other organizations. They also assess the consistency of cultural assessments in two high reliability organizations with Schulman's theoretical typology of high reliability organizations. Additionally, they compare the relationship of culture norms to attitudes and role perceptions in the research with like relationships in high reliability organizations. The data provided in this article show that there appear to be differences between holistic and decomposable high reliability organizations, and they support the Schulman typology.

Keywords: culture, high reliability organization

Knotek, K.; Watson, A.E. 2006. Organizational characteristics that contribute to success in engaging the public to accomplish fuels management at the Wilderness/

non-Wilderness interface. In: Andrews, Patricia L.; Butler, Bret W., comps. Fuels management—How to measure success: Conference Proceedings. 28-30 March 2006; Portland, OR. Proceedings RMRS-P-41. Fort Collins, CO: U.S. Department of Agriculture, Forest Service, Rocky Mountain Research Station: 703-713.

Annotation: See annotation in Section III.C.2. *What is effective crisis communication?* p. 64.

LaPorte, Todd R.; Rochlin, Gene. 1994. A rejoinder to Perrow. Journal of Contingencies and Crisis Management. 2(4): 221-227.

Annotation: In this essay, LaPorte and Rochlin respond to critiques made by Charles Perrow about research completed by the Berkeley High Reliability Organization Project (Berkley group). Specifically, they address five criticisms regarding the Berkley group's research. The criticisms are: 1) the Berkley group holds operational perfection as attainable; 2) the Berkley group perceives high reliability organizations as primarily driven by extreme safety concerns; 3) the Berkley group ignores the complexity in the structural and procedural developments designed to minimize failures and consequences in organizations; 4) the Berkeley group does not find the described phenomena to be atypical or out of the ordinary; and 5) the Berkley group has "gone native" in their research and thereby tainted its results. The authors address each of the criticisms and discuss the future testing and development of high reliability organization theory. In sum, this essay argues that *Normal Accident Theory* (see Perrow 1999 in Section III.A.2.) and *High Reliability Theory* are complementary approaches to understanding organizations that are inherently susceptible to accidents. They note that their work on high reliability organizations attempts to develop a theory of organizational behavior under trying conditions rather than an alternate theory of why or how accidents occur. For a more recent discussion of this debate, see Rijpma (2003, in Section III.B.1.)

Keywords: safety, reliability, high reliability theory, high reliability organizations

Langer, Ellen J. 1997. The power of mindful learning. Reading, MA: Addison Wesley. 167 p.

Annotation: See annotation in Section II.E.2. *How are learning organizations created?* p. 40.

Meyers, Karen K. 2005. A burning desire: assimilation in the fire department. Management Communication Quarterly. 18(3): 344-384.

Annotation: Assimilation and socialization of newcomers prior to entry is more common in high reliability organizations (HRO) than other types of organizations. This study uses Myers and Oetzel's model of organizational assimilation to examine assimilation in an HRO. Newcomers to this HRO tended to develop in their organizational role and become involved with other organizational members. Additionally, in high reliability organizations, emphasis was placed on establishing trust via humility and a strong work ethic. Meyers' study shows the influence of culture and environment on newcomer socialization.

Keywords: organizational assimilation, culture,

Roberts, Karlene H.; Bea, Robert. 2001. Must accidents happen? Lessons from high reliability organizations. Academy of Management Executive. 15(3): 70-79.

Annotation: In recent years, a body of research on high reliability organizations has developed that provides some understanding of disaster and offers advice on how disasters can be minimized in occurrence and severity. Roberts and Bea supplement that body of research by offering advice on how to lessen the likelihood of having an unforeseen accident, regardless of organization type or purpose. They offer three primary suggestions for minimizing error and the potentially hazardous consequences of error. These recommendations are that managers should aggressively seek to know what they do not know, design reward and incentive programs to encourage individuals to recognize both the cost of failure and the benefits of reliability, and strive to communicate the "big picture" to everyone and attempt to get everyone to communicate to one another about how they fit into the "big picture". According to the authors, accidents are unlikely to be eliminated altogether, but by attending to the lessons high reliability organizations offer, organizations may be able to avoid some potential disasters or accidents.

Keywords: management, reliability, disaster

Roberts, Karlene H.; Stout, Suzanne. K.; Halpern, Jennifer J. 1994. Decision dynamics in two high reliability military organizations. Management Science. 40(5): 614-624.

Annotation: This research is part of a larger study examining organizational structure, culture, technology, and decision making in reliability enhancing organizations. Through their work with nuclear-powered aircraft carriers, the authors extrapolate multiple factors that affect the decision making process in two high reliability military organizations. These include a decision process that is constantly in flux; distributed, non-hierarchical decision making; cognitive factors; and task-related factors such as high interdependence. Given the dire consequences associated with faulty decision making in these organizations, this research extends existing organizational theory to more accurately explain decision making in these organizations. The authors provide a model to account for the various changes in the decision making process in these high reliability organizations.

Keywords: decision making, high reliability organization, management, military

Ruchlin, Hirsch S. 2004. The role of leadership in instilling a culture of safety: lessons from the literature. Journal of Healthcare Management. 49(1): 47-58.

Annotation: See annotation in Section II.B.2. *How do cultures impact organizations?* p. 23.

Sagan, Scott. 1995. The limits of safety. Princeton University Press. 302 p.

Annotation: Many organizations tout impeccable safety records, but fail to publicize their flirtations with disaster. In this investigative book, Sagan educates the reader on the untold stories of hazardous technology "close calls". First, he chronicles the origins of accidents, with an emphasis on nuclear weapons safety. The Cuban missile crisis is given special attention, with two chapters dedicated to safety, intelligence, and warning systems during that time. The Thule Bomber accident is also used as a case study to argue that mistakes in high risk arenas are not impossible. The book concludes with a warning that the risk of accidents in high reliability organizations is still disconcertingly high. According to Sagan, in order to learn from these close calls, cover ups and political maneuvering should cease, and our national stance on nuclear positioning should be seriously reconsidered.

Keywords: high reliability organization, accidents, safety, intelligence, politics

Schulman, Paul; Roe, Emery; van Eeten, Michel; de Bruijne, Mark. 2004. High reliability and the management of critical infrastructure. Journal of Contingencies and Crisis Management. 12(1): 14-28.

Annotation: Management of high reliability technical systems and critical service infrastructures has received attention in recent years. However, much of this focus has been on organizations that sustain exclusive control over their technical centers. Notably, many technical systems are not managed by a single organization; instead, they fall under multiple domains. In this essay, the authors address how organizations such as these with competing, if not conflicting, interests and goals sustain highly reliable services, particularly with a sometimes hazy chain-of-command and a rapidly changing environment. The authors argue that increasingly critical infrastructures will consist of interdependent systems, where high reliability will be the expected norm. Finally, the authors discuss the implications for reliability theory and offer practical warnings for critical service infrastructures.

Keywords: reliability, high velocity environment

Weick, Karl; Sutcliffe, Kathleen. 2007. Managing the unexpected: resilient performance in an age of uncertainty. 2nd ed. San Francisco: Jossey-Bass. 224 p.

Annotation: Thousands of firefighters across the United States have been influenced by the first edition of "Managing the Unexpected". In this second edition, the authors continue their analysis of high reliability organizations (HRO's), which are organizations that routinely operate in high risk environments (where the consequences of failure can be fatal) while maintaining extremely low accident/error rates. These low accident/error rates arise from a process of sensemaking that Weick and Sutcliffe call *mindfulness*. The five principles of mindfulness are a *preoccupation with failure*, *reluctance to simplify*, *sensitivity to operations*, *deference to expertise*, and the *ability to be resilient*. HRO's practice mindful organizing and have created mindful infrastructures in which they carry out their dangerous work. The first section of this second edition contains a thorough analysis of the Cerro Grande Fire that burned into Los Alamos, NM, in 2000, causing 250 homes to burn and nearly a billion dollars in damages. Each of the principles of high reliability organizing is discussed in light of the Cerro Grande fire. Other firefighting examples are discussed throughout the book. The principles of HRO and mindfulness that are fully developed in this book would serve as an excellent template for fire managers to begin to think with fresh perspectives about the dangerous fire environments in which they work.

Keywords: high reliability organizing, Cerro Grande, performance, mindfulness, sensemaking

Weick, Karl; Putnam, Ted. 2006. Organizing for mindfulness: eastern wisdom and western knowledge. Journal of Management Inquiry. 15(3): 275-287.

Annotation: In the western world, the concept of mindfulness encompasses the sensemaking attributes which are at the foundation of high reliability organizations (HROs). Examples of HROs include the work operations associated with aircraft carriers, nuclear power plants, hospital emergency rooms, and forest firefighting organizations. In such dangerous high risk occupations, a *mindful* practitioner is able to recognize novel distinctions by being more open to new information, having a greater sensitivity to one's environment, possessing an enhanced awareness of multiple perspectives, and creating new categories on which to structure perceptions. However, according to eastern philosophy, *mindfulness* refers to a mental skill that helps one to stay fully aware in the present moment. Based on eastern wisdom, this mental skill can only be developed during a process of examining one's own mind, which is often accomplished through meditation. In this paper, Weick and Putnam delineate the differences between eastern and western notions of mindfulness. Eastern mental principles of mindfulness are compared to the western perspectives of mindfulness developed by Ellen Langer (see Langer 1997 in Section II.E.2.). The last part of the paper puts mindfulness into perspective from an organizational studies standpoint. Next to Langer's writings, this article may be one of the best explications of how mindfulness can be used as an organizing style of thinking necessary for operating reliably in high risk environments.

Keywords: mindfulness, meditation, training

Weick, Karl E.; Roberts, Karlene H. 1993. Collective mind in organizations: heedful interrelating on flight decks. Administrative Science Quarterly. 38: 357-381.

Annotation: Weick and Roberts attempt to understand why some organizations with a greater possibility for accidents rarely have them. More specifically, they explore the idea that reliable systems are smart systems with more attention placed on reliability than efficiency. The authors contribute fewer accidents to the idea that, because organizations focus on reliability, they are more capable of understanding the complexities they face and responding with fewer errors. In order to test these ideas, Weick and Roberts examine the reliability performance in flight operations of an aircraft carrier. They focused on three ideas: 1) *heed*, or acting with attentiveness, care, and alertness; 2) *conduct*, or behaving while also taking into account others' expectations; and 3) *mind*, or thinking, feeling, and willing. Throughout their analysis, the authors develop the idea of a collective mind that exists in high reliability organizations. *Collective mind* is described as a pattern of heedful interrelations of actions in a social system. The authors find, through their examination of flight operations on aircraft carriers, that as heedful interrelations and comprehension increases, errors decrease within high reliability organizations. The authors conclude the article with theoretical and practical implications.

Keywords: high reliability organization, reliability, collective mind, performance, systems thinking

Weick, Karl. 1991. The vulnerable system: an analysis of the Tenerife air disaster. In: Frost, Peter J.; Moore, Larry F.; Louis, Meryl R.; Lunberg, Craig C.; Martin, Joanne, eds. Reframing organizational culture. Newberry Park, CA: Sage: 117-130.

Annotation: Disasters are often caused by unique combinations of multiple small events. Such combinations are usually unpredictable and complex, making it a challenge to apply lessons learned to future events that are never quite the same. However, an analysis of the Tenerife air disaster reveals how individual and collective stress can transform a relatively calm and disciplined environment into a complex and dangerous one. Pressure causes individuals to fall back on previously learned behaviors and often makes it more difficult for them to understand and effectively respond to complex situations. As communication is distorted and/or lost within an organization, the organization itself becomes weak, the chain of command fades away, and members begin acting in more individualistic ways. Leaders who understand how people typically react in high stress situations, and who put a priority on maintaining clear communication and social interaction, are more likely to anticipate and prevent potential disasters.

Keywords: high reliability organization, Tenerife air disaster, disaster, communication, stress, disaster prevention, systems thinking

C. Crisis Communication

Crisis communication refers to how organizations manage their identity and relationship with the public before, during, and after a crisis. Often, an organization's success or failure depends on how effective its communication is during a crisis. This section begins by defining crisis communication in terms of public relations appeals to both internal and external audiences (Seeger and Ulmer 2002; Seeger and Ulmer 2003). This section includes a number of real life crises and how they were managed by different communication strategies, such as retrospective sensemaking (Sellnow and Seeger 2001) and strategic ambiguity (Ulmer and Sellnow 1997).

The next set of readings discusses how to communicate effectively during a crisis. This section includes general tips (Schwarz and others 2007), applies the concepts of mindfulness to communication (Knotek and Watson 2006), and explores studies of specific organizations, such as Enron and Nike, that have been damaged by image attacks and have made communication mistakes (Conrad 2003; Knight and Greenberg 2002). This section also discusses how organizations have effectively avoided and/or prevented image problems before a crisis occurred. Although a preemptive solution to a crisis is always the most desirable, some authors examine how to fix image problems when a crisis situation is unavoidable (Benoit 1997; Seeger and Ulmer 2002). Further readings can be found in Arvai and others (2004).

Author's Picks

- Arvai, J.; Wilson, R.; Rivers, L.; Froschauer, A. 2004. Environmental risk communication: a review and annotated bibliography. Unpublished report. Eugene, OR: Decision Research. 165 p. [Note: This publication is available at the USDA Forest Service's Pacific Southwest Research Station via email (pwinter@fs.fed.us) or phone (951-680-1557).

- Coombs, W. Timothy. 1999. Ongoing crisis communication: Planning, managing, and responding. Thousand Oaks, CA: Sage. 192 p.

- Fink, Steven. 2002. Crisis management: planning for the inevitable. Lincoln, NE: iUniverse, Inc. 264 p.

- Seeger, Matthew W.; Ulmer, Robert R. 2003. Explaining Enron: Communication and responsible leadership. Management Communication Quarterly. 17(1): 58-84.

1. What is Crisis Communication?

Arvai, J.; Wilson, R.; Rivers, L.; Froschauer, A. 2004. Environmental risk communication: a review and annotated bibliography. Unpublished report. Eugene,

OR: Decision Research. 165 p. [Note: This publication is available at the USDA Forest Service's Pacific Southwest Research Station via email (pwinter@fs.fed.us) or phone (951-680-1557).

Annotation: This bibliography was compiled to "provide a comprehensive analysis of these literature bases and provide the USDA Forest Service with a detailed annotated bibliography that covers seminal works" on risk (including risk assessment, risk characterization, risk communication, risk management), stakeholder involvement, and decision science. The annotated bibliography reviews 152 articles the authors deemed useful for designing decision-focused risk communication processes. Readings include three types of journal articles (research articles, review articles, and case studies), books and book chapters, and agency reports. Following a brief introduction to the topic of risk communication, the authors provide a one-page synopsis for each included reading.

Keywords: risk communication, risk management, stakeholders, decision making

Coombs, W. Timothy. 1999. Ongoing crisis communication: planning, managing, and responding. Thousand Oaks, CA: Sage. 192 p.

Annotation: After reviewing the crisis management literature, Coombs argues that the study of crisis management is fragmented in two ways: multidisciplinary work is often eschewed and researchers tend to study only part of a crisis. In this book, Coombs seeks to provide an integrative and holistic view on crisis management. He advocates a three-stage model of crisis communication. The first stage, or *precrisis stage*, includes detecting, preventing, and preparing for crises. The second stage, or *crisis event*, includes recognizing, containing, and recovering from the crisis. Finally, the *postcrisis stage*, includes evaluation, reflection, and identification of areas for future action. Coombs defines and describes each stage in succession and provides a number of "application points" and case studies. His writing is solidly grounded in academic literature on crisis management and pays particular attention to issues management and stakeholder concerns in each stage of a crisis.

Keywords: crisis communication, crisis management, stakeholders

Fink, Steven. 2002. Crisis management: planning for the inevitable. Lincoln, NE: iUniverse, Inc. 264 p.

Annotation: When he was unexpectedly drafted onto the Three Mile Island crisis management team in 1979, Steven Fink received a firsthand education in the challenges of crisis management. This knowledge, and his experience since then, forms the basis of this practical guide to crisis management. Fink argues that a crisis is like a disease with four distinct phases. The first phase, or *prodomal phase*, is a warning phase before a crisis occurs. Fink argues that an employee's job is to sharpen the individual's senses to detect warnings early. The second phase of a crisis is the *acute crisis phase*, the phase where the most damage is incurred. During this phase, an employee must try to control the crisis as much as possible. Following the acute crisis phase is the *chronic crisis stage*, where cleanup, recovery, investigation, and healing might happen. This phase allows managers to assess how the crisis was managed and adjust future crisis management plans. Finally, the fourth phase of a crisis is the *crisis resolution stage*, where business returns to normal. Fink argues that, because crisis is inevitable, all organizations should have a comprehensive crisis management plan. Fink's practical advice for communicating during a crisis situation is bolstered by a number of real life examples, such as the Three Mile Island and Tylenol crises.

Keywords: crisis communication, crisis management, public relations

Sellnow, Timothy L.; Seeger, Matthew W. 2001. Exploring the boundaries of crisis communication: the case of the 1997 Red River Valley flood. Communication Studies. 52(2): 153-168.

Annotation: Sellnow and Seeger use the 1997 Red River Valley flood in North Dakota as a case study to explore some of the current crisis communication literature. They examine flood communication in four specific ways: 1) chaos theory, 2) retrospective sensemaking, 3) crisis communication logistics, and 4) corrective actions. First, *chaos theory* illuminated the need to be flexible. Routine procedures that had been used to communicate prior to the crisis did not account for a variety of situations that might have impacted the situation. Second, the authors consider how *retrospective sensemaking* led to a lack of flexibility during the 1997 floods, as managers were unable to see outside of a particular framework of how floods had previously occurred. Third, the authors explore the ways in which specific and immediate *crisis communication* practices needed to be in place so residents could easily obtain information during the flood. Further, they look at how the use of a particular newspaper allowed residents to stay in touch with their community, even in the event of an evacuation. Finally, the authors look at how consequent *corrective actions*, in a variety of agencies and institutions, were used to begin rebuilding the community after the flood. This article illuminates the variety of ways in which crises can be viewed as influences of subsequent actions that might help avoid similar crises later.

Keywords: crisis management, crisis communication

Sellnow, Timothy L.; Seeger, Matthew W.; Ulmer, Robert R. 2002. Chaos theory, informational needs, and natural disasters. The Journal of Applied Communication Research. 30(4): 269-292.

Annotation: The authors apply chaos theory to communication during the 1997 Red River flood in Minnesota and North Dakota. *Chaos theory* looks at the ways in which large complex systems move back and forth between order and chaos. The authors examine the conditions and decisions leading to the disaster and consequent reactions to the crisis. Communicative behaviors are analyzed four ways. First, *fractals* are complex, but "self-similar" patterns that appear in seemingly chaotic systems. In other words, while there appears to be a degree of variability, there are still characteristics that are similar within comparable systems. Second, cosmology episodes are occurrences that make people believe that there is chaos within the overall system. The people involved are overwhelmed with the situation itself. *Cosmology episodes* are points where actors realize that their actions are novel. Third, *self-organization* occurs within chaotic situations. The authors describe it as a natural process by which order comes out of a chaotic state. The order comes from an inner sense of values rather than order imposed upon the system by an external force. Fourth, the concept of *strange attractor* is one where order and patterns evolve out of chaos and help to return a situation to some sort of balance. Thus, the authors analyze communication behaviors related to: 1) river crest predictions (fractals), 2) the shock of the magnitude (cosmology episode), 3) novel ways of reorganizing (self-organization), and 4) agencies that aided in establishing a new order (strange attractors). The authors conclude by identifying research needs, but also emphasize the fact that crises are a part of "business as usual" that help organizations grow and renew themselves, as well as reinforce possibilities for change.

Keywords: chaos theory, crisis management, systems thinking

Ulmer, Robert R.; Sellnow, Timothy L. 2002. Crisis management and the discourse of renewal: understanding the potential for positive outcomes of crisis. Public Relations Review. 28(4): 361-365.

Annotation: This short article examines crisis communication that emphasizes renewal and growth rather than blame, responsibility, and liability. Three categories of renewal are discussed in light of the September 11, 2001, terrorist attacks. These categories include renewal based upon stakeholder commitment, commitment to correction, and core values. The authors conclude with implications for crisis communication focused on renewal and the positive outcomes. Ulmer and Sellnow note how crisis communication, when based in an organization that has good will toward those who are benefited by the organization, can lead to renewal and growth.

Keywords: crisis communication, 9/11

Ulmer, Robert; Sellnow, Timothy. 1997. Strategic ambiguity and the ethic of significant choice in the tobacco industry's crisis communication. Communication Studies. 48: 215-233.

Annotation: Ulmer and Sellnow address how the concept of strategic ambiguity can be used during a crisis to influence interpretations of an organizational crisis. Strategic ambiguity refers to when an organization chooses to be vague or indistinct about particular situations when there are multiple possible interpretations of events. The article uses the tobacco industry as a case study of ways that strategic ambiguity can undermine the credibility of the organization. The authors conclude that strategic ambiguity can be misused and detrimental to the organization using it.

Keywords: crisis management, crisis communication, strategic ambiguity

Williams, David; Treadaway, Glenda. 1992. Exxon and the Valdez accident: a failure in crisis communication. Communication Studies. 43: 56-64.

Annotation: The authors use the Exxon Valdez oil tanker accident as a case study to argue that Exxon's overall crisis management efforts were a failure due to their slow response. Rather than issuing a "strong initial statement immediately following the accident," their response was reactive. Further, rather than attempting to limit the environmental impact of the accident immediately, their attempts were delayed. Williams also argues that Exxon unsuccessfully used two crisis management strategies, *burden sharing* and *scapegoating*. *Burden sharing* refers to Exxon's attempts to sidestep reasons for their lack of immediate response to the crisis. *Scapegoating* refers to attempts to place direct blame on someone (or something) else for the crisis. Examining ways in which an organization failed to manage crisis effectively helps to see what does not work and highlights issues to take into consideration when faced with crisis communication.

Keywords: crisis management

2. What is Effective Crisis Communication?

Benoit, William L. 1997. Image repair discourse and crisis communication. Public Relations Review. 23(2): 177-186.

Annotation: Maintaining a productive and positive image is essential for both individual and organizational credibility. Attacks from public or internal parties can damage individuals and organizations in ways that are difficult to repair. However, individuals and organizations have several ways to repair images after they have been attacked and/or damaged. According to the author, an image attack is made up of two elements. First, the person or organization is perceived as the party responsible for some kind of action. Second, that action is considered offensive. For those who must deal with image attacks, he recommends five image repair techniques: denial, evasion of responsibility, reducing offensiveness of event, corrective action, and mortification. Following an explanation and illustration of each technique, he makes five recommendations for effective image repair communication, including acknowledgement of responsibility when the responsibility is legitimate, acknowledgement of factors that are beyond one's control, reporting plans to correct and/or prevent recurrence of the problem, using multiple strategies/techniques simultaneously, and recognizing the limitations of persuasive image repair techniques. These techniques can be used to prepare for, and manage, image attacks when they occur.

Keywords: image, image repair, communication

Conrad, Charles. 2003. Setting the stage: introduction to the special issue on "corporate meltdown". Management Communication Quarterly. 17(1): 5-19.

Annotation: While business leaders and employees must be held accountable for the breakdown of ethics leading up to crises such as Enron, such choices cannot be viewed in a vacuum. Conrad argues that three economic and social influences played a major role in shaping the actions of the individuals involved in this disaster: free market fundamentalism, a discourse of a "new economy," and the CEO as a secular savior. Managers can use this example by keeping in mind the economic and cultural elements that worked together to produce unethical behavior. This article suggests that managers must weigh personal accountability within the context of cultural and societal influences.

Keywords: culture, ethics

Knight, Graham; Greenberg, Josh. 2002. Promotionalism and subpolitics: Nike and its labor critics. Management Communication Quarterly. 15(4): 541-570.

Annotation: Nike's use of symbolic communication to build and maintain their corporate image has helped establish a worldwide identity. However, their prominence through this promotionalism has also made them the target of critical image attacks, calling into question their credibility as an organization. This article provides a case study of how one organization responded to image attacks and the challenges they faced in doing so.

Keywords: image, communication, image attack

Knotek, K.; Watson, A.E. 2006. Organizational characteristics that contribute to success in engaging the public to accomplish fuels management at the Wilderness/non-Wilderness interface. In: Andrews, Patricia L.; Butler, Bret W., comps. Fuels management— How to measure success: Conference Proceedings. 28-30 March 2006; Portland, OR. Proceedings RMRS-P-41. Fort Collins, CO: U.S. Department of Agriculture, Forest Service, Rocky Mountain Research Station: 703-713.

Annotation: Using a prescribed burn on the Lewis and Clark National Forest in Montana as a case study, Knotek and Watson applied Weick and Sutcliffe's model (see Weick and Sutcliffe 2007 in Section III.B.2.) of mindfulness and high reliability organizing to the public outreach efforts needed to plan and ignite a successful prescribed burn. Their central thesis was: could the five basic tenets of mindfulness—preoccupation with failure, reluctance to simplify, deference to expertise, resilience, and sensitivity to operations—be used to identify and understand the characteristics of an organization that contribute to a successful interaction with the public about a proposed hazardous fuel treatment project? To test their hypothesis, the authors interviewed 14 local agency representatives and 24 local non-agency public representatives. Using the transcripts from these interviews, the authors qualitatively screened the results through the five tenets of mindfulness. Their results showed that the five mindfulness processes associated with high reliability organizing were a useful tool for identifying, categorizing, and describing public outreach efforts in terms of their mindfulness. This research showed that the public outreach efforts associated with the case study burn were generally mindful, and, in all but one case, the public's perceptions of the agency's ability to respond and recover from an unexpected event could be easily identified. The authors conclude with suggestions on how a framework of mindfulness might be applied to public outreach efforts. The approach outlined in this paper to using mindfulness is one of the first instances where mindfulness has been used to analyze a fire management problem not associated with firefighting safety. Knotek and Watson's research demonstrates that there are tremendous possibilities in applying high reliability theory to other natural resource issues and that mindfulness is a concept that may be useful in other work activities to help make better sense of everyday work worlds.

Keywords: mindfulness, public outreach

MacGregor, Donald G.; Finucane, Melissa; González-Cabán, Armando. 2007. The effects of risk perception and adaptation on health and safety interventions. In: Martin, Wade E.; Raish, Carol; Kent, Brian, eds. Wildfire Risk: Human Perceptions and Management Implications. Washington, DC: Resources for the Future Press: 142-155.

Annotation: This book chapter aspires to help managers reach communities and homeowners about ways to reduce their risk of loss due to wildland fire (for example, voluntarily removal of flammable materials near buildings). The authors begin by defining an *intervention* in this context as a purposive effort by an agency charged with protecting citizens' safety, to change citizen knowledge, attitudes, and/or behavior in order to reduce the economic and/or social cost of wildland fire. Examples of interventions include media campaigns, town meetings, brochures, and workshops designed to promote awareness and understanding of the risk. The authors explain various factors that influence the public's response to an intervention. These include cognitive and emotional factors (for example, fire risk knowledge, personal risk perception, decision skills, and financial incentive), as well as social and cultural factors (for example, social and cultural values, information delivery format, and availability of resources).

Next, the authors offer a conceptual framework, the *Social Amplification of Risk Framework (SARF)*, to explain how people interpret risk events and how these interpretations spread from individuals to organizations and, ultimately, to larger social enterprises. In this framework, people interpret and assign meaning to risk events based on the interaction of a variety of individual and social agents. They also interpret perceptions and attitudes they hold for the organizations associated with risk management. SARF is especially helpful in understanding social reactions to risk that seem stronger than warranted based on technical criteria. The framework can also be used to understand how people adapt to risk events; that is, how they assimilate risk-related events into pre-existing cognitive and/or emotional structures and how they modify or reorganize risk-related attitudes, perceptions, and behavior based on their experiences. The authors conclude by defining interventions as the risk events that can lead to changed perceptions of risk and adaptive behavior. They note that risk-related behavior is adaptive, interventions related to wildland fire should be both long-term and specific to targeted populations, and further study is needed to evaluate the impact of interventions on behavior change. This information is also applicable to outreach efforts for fuel reduction projects (for example, prescribed fire, and brush or tree removal) in the wildland-urban interface.

Keywords: risk adaptation, risk communication, behavior, public outreach

Schwarz, Roger; Davidson, Anne; Carlson, Peg; McKinney, Sue. 2005. The skilled facilitator fieldbook: tips, tools and tested methods for consultants, facilitators, managers, trainers and coaches. San Francisco, CA: Jossey-Bass. 576 p.

Annotation: This fieldbook, a compendium of 62 articles by consultants, facilitators and organizational psychologists, is organized into seven parts that describe in detail the *skilled facilitator* approach to running meetings and facilitating groups. This approach seeks to understand the core values and assumptions working within groups and to rigorously show how these values and assumptions increase or decrease a group's overall effectiveness. Concepts such as effective facilitation, group intervention, how mental models clash (my picture of the world is not the same as yours), theories of action (what we espouse compared to what we actually do), ladder of inference (how do my assumptions and inferences affect what I'm thinking and doing?), systems thinking (how all the parts of the system fit together), and the left hand column exercises (what is not being said in a conversation compared to what is being said), are described in detail. Much of the philosophy of the skilled facilitator technique is grounded in the "theory of action" studies completed by Chris Argyris (see Argyris 1990 Section II.E.1.) and is relevant to developing a learning organization. The authors conclude with discussions on deepening one's practice, the challenges and risks of applying the skilled facilitator techniques, and how to apply those same processes as a leader of change within an organization and in one's non-work life. This is an excellent manual for anyone facilitating a meeting, teaching a class, or leading a discussion on a complex issue. It is also an excellent guidebook to becoming a better facilitator.

Keywords: facilitation, systems thinking, leadership, organizational learning, values, groups, theory of action

Seeger, Matthew W.; Ulmer, Robert R. 2003. Explaining Enron: communication and responsible leadership. Management Communication Quarterly. 17(1): 58-84.

Annotation: Enron's sudden and shocking fall from economic power has resulted in dozens of explanations. However, unlike previous explanations, Seeger and Ulmer argue that Enron's demise was fundamentally a result of leadership flaws in communication-based responsibility. They recognize three areas of ethical communication that Enron's leaders failed to live up to: communicating appropriate values to create a moral climate, maintaining adequate communication to be informed of organizational operations, and maintaining openness to signs of problems. After explaining each flaw in detail, they make three recommendations for managers. First, leaders are obligated to not only talk about ethical behavior, but to demonstrate it. Second, leaders are responsible for knowing what is happening within their organizations despite their areas of expertise. Third, leaders must be open to bad news, dissent, warnings, and problem signs and are responsible for nurturing an environment that encourages such expression.

Keywords: communication, leadership, ethics

Seeger, Matthew W.; Ulmer, Robert, R. 2002. A postcrisis discourse of renewal: the cases of Malden Mills and Cole Hardwoods. Journal of Applied Communication Research. 30(2): 126-142.

Annotation: Organizations have typically looked at crisis situations as something negative that requires some form of apology to the public or other related groups. However, Seeger and Ulmer argue that more productive and positive alternatives are available. In reviewing these types of situations, they present two case studies of organizations that were faced with similar disasters. Three themes emerge from an analysis of their responses: commitment to stakeholders, commitment to rebuild, and organizational renewal. Several sets of practical implications leading to greater success during crisis situations are given. These include a positive pre-crisis relationship with employees and other stakeholders, quick response time to the crisis, and a mindset that a crisis can often be a positive situation that provides opportunities for organizational renewal and growth.

Keywords: organizations, crisis, crisis management, disaster

Ulmer, Robert R. 2001. Effective crisis management through established stakeholder relationships: Malden Mills as a case study. Management Communication Quarterly. 14(41): 590-615.

Annotation: Ulmer examines the post-crisis communication used by the chief executive officer of Malden Mills after there was a fire at a textile mill. Ulmer finds that having strong and effective channels of communication, as well as specific value positions with stakeholders prior to crises, helps in post-crisis communication. The recovery process can be eased by having these communication channels in place prior to a crisis. Further, Ulmer finds that if the organization is able to use all levels of crisis communication management, there can be positive outcomes such as positive media attention, stakeholder advocacy, and instrumental communication with stakeholders.

Keywords: crisis management, stakeholders

IV. Internet Resources

These online resources provide information and tools related to wildland fire safety, communication, leadership, and mindfulness. We hope this combination of government and non-government Internet sites is useful to readers; however, the inclusion of non-government sites does not imply endorsement of their content. These websites were active as of July 6, 2007.

A. Safety and Training

These websites provide information to prepare for and tactically manage wildland and prescribed fire incidents. This includes information about local incidents; situation reports; weather, fuels, and fire danger predictions; software applications; resource coordination; maps; fire policy; national standards; the Standard Fire Orders, Watch Out Situations, and LCES (Lookouts, Communication, Escape routes, Safety zones); leadership tools and tips; safety alerts; aviation safety; the Wildland Fire Safety & Health Reporting Network (SAFENET); the Job Hazard Analysis (JHA) database; and the Incident Management Team Center (IMTcenter.net). While the information on these sites is primarily geared toward tactical operations, it also includes fire statistics, lessons learned, research (for example, the Tridata Safety Awareness Study), training resources, fatality reports, investigations, and accident reviews. Most of these sites are hosted in the United States; however, there are websites from Australia and Canada, which link to additional international sites.

Australian Capital Territory — Fire Safety

Available: http://www.esb.act.gov.au/Fire_Safety

For residents in the Australian Capital Territory (ACT), this site offers a wealth of resources. Links are organized by what to do before, during, and after a bushfire. The site links to the Strategic Bushfire Management Plan, Bush Firewise, and the publication "Bushfires and the Bush Capital," which helps householders identify risks, prepare their properties, and know what to do in the event of a bushfire. Additional links include a map for use when reporting the location and extent of bushfires and other emergencies in the ACT, weather and other links helpful for risk assessment, media releases, the Fire Brigade's community safety bulletins, and the Rural Fire Service.

Bureau of Land Management — Fire and Aviation Safety

Available: http://www.fire.blm.gov/WhatWeDo/safety.htm

To provide resources for Bureau of Land Management firefighters, this site links to national standards for fire operations (the Red Book); fire preparedness reviews; the National Interagency Fire Center's safety resources; firefighter pocket guides; and information on Standard Fire Orders, Watch Out Situations, and LCES (Lookouts, Communication, Escape routes, Safety zones).

Canadian Forest Service

Available: http://fire.cfs.nrcan.gc.ca/index_e.php

Hosted by Natural Resources Canada, Canadian Forest Service, this site links to daily fire maps, weekly fire statistics, fire research, FireSmart resources to protect communities, the Canadian Wildland Fire Strategy, a large fire database, historical weather and fire behavior analyses, general facts about forest fires in Canada, a searchable database of publications, and a long list of Canadian and International fire links.

Fire Leadership: Developing Leaders in Wildland Fire

Available: http://www.fireleadership.gov

Administered by the National Wildfire Coordinating Group's Leadership Committee, this website hosts a variety of resources for managers who want to improve leadership skills at all career levels. It includes links to workshops, wildland fire values and principles, training courses, goal setting tutorials, and regular announcements of new features offered on the website as well as upcoming events in the fire community.

Fire Leadership Toolbox

Available: http://www.fireleadership.gov/toolbox/toolbox.html

The Fire Leadership Toolbox links to the professional reading program, online courses for leadership skills, a staff ride guide and library (including Cerro Grande, South Canyon, and Thirtymile staff rides), a tactical decision games workbook and library, examples of sand tables, standard operating procedures workbook, the crew cohesion assessment, tips for briefings, leadership reaction scenarios, an after action review training package, a self development plan, and biographies of "leaders we would like to meet".

Geographic Area Coordination Centers

Available: http://gacc.nifc.gov

Primarily for use by local and geographic area wildland fire managers and firefighters, Geographic Area Coordination Centers (GACC) provide interagency logistical coordination for managing wildland fires and mobilizing firefighting resources. Each GACC is responsible for coordinating resource mobilization between the units within the Geographic Area and providing predictive services and intelligence products for decision support. In addition to local incident, logistics, and dispatch information, these sites link to weather, fuels, and fire danger prediction sites, situation reports, fire policy, maps and imagery, software applications, training, safety resources, and fire statistics.

National Advanced Fire & Resource Institute

Available: http://www.nafri.gov

This site provides basic information about the National Advanced Fire & Resource Institute. The site lists course offerings, explains the nomination process, and provides other information for wildland firefighters who want to become qualified to manage wildland fires on a regional level.

National Business Center Aviation Management— Aviation Safety & Evaluation Division

Available: http://amd.nbc.gov/safety

Primarily serving the Department of the Interior, the Aviation Management Directorate's goals are "...to raise the safety standards, increase the efficiency, and promote the economical operation of aircraft activities in the Department of the Interior". Dealing specifically with aviator safety in the field of wildland fire, this site offers links to publications, training/education, safety alerts, the Aviation Safety Communiqué mishap reporting database (SAFECOM), aviation accident prevention bulletins, the Interagency Aviation Mishap Response Guide & Checklist, and aviation safety accident reviews.

National Firefighter Fatality Investigation and Prevention Program

Available: http://www.cdc.gov/niosh/fire

This site is hosted by the National Institute for Occupational Safety and Health (NIOSH), which conducts independent investigations of both structural and wildland firefighter line-of-duty fatalities. The website provides access to the NIOSH investigation reports, which can be queried by state, year, or medical- or trauma-related category. Wildland fire traumas are separated into burnovers, urban/interface, prescribed burns, motor vehicles, and other. The site also links to biweekly firefighter safety quizzes, many of which are structure related. There is a 2003 report that addresses deficiencies in firefighter radio communications and identifies emerging technologies that may address these deficiencies.

National Interagency Fire Center—Safety

Available: http://www.nifc.gov/safety_study

This extension of the National Interagency Fire Center's website provides links to materials on training and education, safety communication, and SAFENET. It also hosts links to historical wildland fire fatality reports, the Job Hazard Analysis (JHA) Database, the Federal Fire and Aviation Safety Team, the National Institute for Occupational Safety and Health (NIOSH), the Department of the Interior's SafetyNet, and many others.

National Wildfire Coordinating Group's Safety and Health Working Team

Available: http://www.nwcg.gov/teams/shwt/index2.htm

Subscribing to the values embodied in high reliability organizations, the National Wildfire Coordinating Group's Safety and Health Working Team strives to "improve firefighter health, safety, and effectiveness through emphasis on excellence in workforce development, leadership, and the establishment of standards". The team collects and analyzes data, and prioritizes safety issues for resolution and communication to the field and management. This site links to a reference manual, team activities, safety alerts, information on entrapments and fatalities, the hazard tree/tree felling task group, federal agency safety links, national and international safety organizations, and educational and research sites.

Wildland Fire Safety & Health Reporting Network (SAFENET)

Available: http://safenet.nifc.gov

The development of SAFENET was recommended in Phase III of Tridata's Wildland Firefighter Safety Awareness Study (see Tridata 1998 in Section I.). The program's objectives are to provide a forum for firefighters to voice their safety concerns, facilitate problem solving, and aid in identifying trends as they relate to firefighter safety. Anyone can anonymously fill out a SAFENET, anytime, to report a valid concern about unsafe situations in wildland fire, prescribed fire, or all risk operations. This site allows SAFENET reports to be submitted and viewed, offers help on completing the form, describes the SAFENET protocol and administrative process, answers frequently asked questions, and links to annual SAFENET reviews. SAFENET is managed by the Federal Fire and Aviation Safety Team through direction from the NWCG Safety and Health Working Team.

USDA Forest Service, Fire & Aviation Management— Wildland Fire Safety

Available: http://www.fs.fed.us/fire/safety

This page hosts a directory of fire safety links, including alerts and advisories, the Code of Federal Regulations, fire shelter information, fitness publications/brochures, the health and safety handbook and reports, investigations, the

Safety Awareness Study (Tridata 1996a, 1996b, 1998 see annotations in Section I.), the Wildland Fire Safety and Health Network (SAFENET) (see annotation in this section), Safety Zone newsletters, training, shelter deployment reports, radio communications, and more.

Wildland Fire Lessons Learned Center

Available: http://www.wildfirelessons.net

The Wildland Fire Lessons Learned Center (LLC) website is an interagency site focused on improving safe work performance through organizational learning in wildland fire organizations. Designed for the entire incident management community, this site hosts lessons learned, effective practices, near miss reports, incident reviews, and links to a variety of other websites dealing with wildland fire training and leadership. It hosts "Scratchline," a quarterly newsletter sharing lessons on tactics, techniques, procedures, and processes that are identified from the field through After Action Review Rollups and Information Collection Team interviews. The site links directly to MyFireCommunity. net, which is an online community center designed to assist wildland fire work groups in identifying one another, sharing learning opportunities, discussing issues and concerns, and exchanging knowledge. The LLC site also links to the Incident Management Team Center (IMTcenter.net), which enables teams to set up and maintain their own websites without the need for a webmaster or special software training.

B. Research, Theory, and Management

These websites provide current and emergent information related to creating and maintaining an active learning environment. They offer practical tips, summarize and provide access to the latest publications, present brief descriptions of the theoretical concepts behind learning organizations, and offer opportunities to network with others who are pursuing and applying knowledge in these arenas. There are links to a wildland fire professional association and wildland fire publications that cover a variety of topics aimed at understanding and improving wildland fire management. In addition to the theory of action and organizational learning, these sites provide readings on wildland fire leadership, wildland fire decision processes, how to apply high reliability organizing principles in wildland fire, and other social research conducted on wildland fire organizations and their interactions with the public. Some of these sites supply bibliographies and/or libraries for those interested in obtaining a deeper understanding of these concepts and how to use them.

Action Design Resources

Available: http://www.actiondesign.com/resources

This site hosts a variety of resources for those interested in the theory of action and action science as described by Chris

Argyris and Donald Schön. The site has a bibliography, including suggestions of "what to read first," a description of the core concepts of action science, guidelines for how to write a case study to learn from an actual situation, tips on how to start a learning group, and the full text of the 1985 book "Action Science" by Chris Argyris and others.

Fieldbook.com

Available: http://www.fieldbook.com

Hosted by the non-profit Society for Organizational Learning, this is the site of The Fifth Discipline Fieldbook Project. The site is intended to "serve the far-flung community of learning organization practitioners. These are people drawn together by the idea of a learning organization". The site describes the five "learning" disciplines codified in "The Fifth Discipline" (see Senge 1990 in Section II.E.1.). Through this site, readers can complete a registration form to stay connected with new developments, tools, and organizational learning resources as they emerge; learn about and order the following books by Senge and others: "The Dance of Change, Schools that Learn," and the "Fifth Discipline Fieldbook" (see Senge and others 1994 in Section II.E.2.); read new material from the authors and contributors of these books; and provide advice or query others for help and advice through the Bulletin Board. There are links to organizational learning events on topics such as leadership, systems thinking, and organizational change; notes on how to form a study group; and geographically organized lists of people who want to join or start groups.

Fire Leadership Professional Reading Program

Available: http://www.fireleadership.gov/toolbox/documents/pro_reading_room.htm

The Fire Leadership Professional Reading Program provides "a selection of readings that will support continuing education efforts within the wildland fire service". It links to both online and hard copy versions of the professional reading list, offers suggestions for developing a local reading program, and provides a place to suggest additional readings. The professional reading list includes readings on fire history and culture, human factors, lessons learned, leadership and management, case studies on leadership, and books recommended by the United States Federal Agency National Fire Directors.

Fire Management Today

Available: http://www.fs.fed.us/fire/fmt

"Fire Management Today" is one of the wildland fire community's longest running publications. With an underlying theme of wildland fire safety, it has served as a knowledge base for new techniques, issues, and industry trends for 70 years. This website hosts a searchable database of back issues from 1997 to present and provides contact information to obtain access to earlier issues.

Foundations of Action Design

Available: http://www.actiondesign.com/resources/theory

This site offers a brief description of the theory and practice upon which action science is based. It covers the theory of action (Chris Argyris and Donald Schön), family systems (David Kantor), developmental theory (Robert Kegan), and organizational learning (Peter Senge and others). Those interested in additional detail on action science can find it through many additional links.

International Association of Wildland Fire

Available: http://www.iawfonline.org

The International Association of Wildland Fire (IAWF) is a non-profit, professional association that works toward facilitating communication and providing leadership for the wildland fire community. The site links to the association's two publications, "Wildfire Magazine" and the "International Journal of Wildland Fire". It also links to IAWF newsletters, news releases, position papers, and awards; "Wildfire Magazine" editorials; conference information; the Amazon.com Associate Program on wildfire books; the Wildland Fire Event Calendar; and brief descriptions of the multiple fatality wildland fires around the world over the last 150 years that are included in the Wildland Fire Event Calendar (the intent here is that the lessons learned from past fires will be less likely to be forgotten).

Leopold Institute—Wilderness Fire Research

Available: http://leopold.wilderness.net/research/fire.htm

One of this fire research program's three objectives is to "understand the social and institutional factors that can affect how managers and the public make sense of, and decisions about, fire management". The program's social research focuses on: 1) understanding the constraints to managing fire at landscape scales, 2) deepening knowledge of decision-processes and sensemaking as they influence fire operations and outcomes, 3) improving methods for assessing the social consequences of fire management activities, and 4) understanding public knowledge, attitudes, and behaviors related to fire management in and surrounding wilderness. The site describes research priorities, links to project descriptions and products, and includes a variety of resources for those interested in understanding, from a research perspective, the social side of wilderness fire management.

Pacific Southwest Research Station—Social Aspects of Fire

Available: http://www.fs.fed.us/psw/topics/recreation/fire

This site summarizes the Pacific Southwest Research Station's social science research that is relevant to fire management. Their main research initiatives are focused on recreation trends, communication techniques, social impacts of fire and fire management at the urban-wildland interface, and behaviors and conflict (in other words, attitudes, values, behaviors, conflict, recreation participation, and decision making). The site describes research projects and the associated products.

Paul Gleason and LCES Worldwide

Available: http://www.wildlandfire.com/docs/gleason/links.htm

This website hosts a variety of documents that look at Paul Gleason's history as a leader in wildland firefighting and the worldwide impact of the Lookouts, Communication, Escape Routes, Safety Zones (LCES) system. The site links to the USDA Forest Service's LCES web page.

Society for Organizational Learning

Available: http://www.solonline.org

This non-profit, member-governed corporation evolved from the Massachusetts Institute of Technology's (MIT) Center for Organizational Learning. Peter Senge, author of the "The Fifth Discipline: the Art and Practice of the Learning Organization" is the founding Chairman of the Society for Organizational Learning. The site offers participation in a learning organization community, consulting, coaching, courses and programs, and publishing. The site describes the five disciplines of organizational learning as described by Senge, offers links to publications and resources, and provides a long list of learning communities, related websites, and a bibliography.

Thoughts on Leadership

Available: http://www.guidancegroup.org/library.php

This page provides text from essays originally published in the "Wildfire Magazine" column *Thoughts on Leadership*. Articles are generally focused on leadership skills, strategic organizational planning strategies, and building community/organizational collaboration. Specific topics include risk taking, leadership vs. management, trust and teamwork, accountability and responsibility, multi-frame thinking, vision, power, and openness/transparency. This site is hosted by Guidance Group, Inc., a private consulting firm focused on fire service organizations.

Wildland Fire Lessons Learned Center: High Reliability Organizing

Available: http://www.wildfirelessons.net/HRO.aspx

This page summarizes recent efforts and products that promote the implementation of high reliability organizing (HRO) within the wildland fire community. Resources include workshops, Powerpoint presentations, HRO assessments, summaries of past "Managing the Unexpected" workshops, an HRO incident organizer, and a worksheet exercise on Taking HRO Home.

Wildland Fire Lessons Learned Center Library

Available: http://www.wildfirelessons.net/Library.aspx

Visitors to this page have access to an international database of literature dealing with a seemingly limitless number of wildland firefighting topics. Site visitors can submit documents to the library, browse existing documents by topic or media type, or search the library's contents. Media types include incident reviews, incident toolboxes, incident collection team reports, other reports and journal articles, and after action reviews. The database is constantly growing and changing and is a resource to those interested in safety and lessons learned.

C. Blending Eastern and Western Notions of Mindfulness

The following websites have been recommended by Ted Putnam (ex-smokejumper, fire fatality investigator, experimental psychologist, and organizer of the 1995 Human Factors Workshop) for those interested in pursuing more information on the concepts presented in Weick and Putnam (2006, see annotation in Section III.B.2.). The Weick and Putnam paper compared eastern principles of mindfulness (evident in eastern religions such as Buddhism) with the western notion of mindfulness as described by Langer (1997, see annotation in Section II.E.2.). These websites offer resources developed by bringing western scientists and eastern leaders (such as the Dalai Lama) together to better understand how the mind and brain function.

Center for Mindfulness in Medicine, Health Care and Society

Available: http://www.umassmed.edu/cfm/index.aspx

Based at the University of Massachusetts Medical School, this Center is dedicated to "furthering the practice and integration of mindfulness in the lives of individuals, institutions, and society" through clinical, research, education, and outreach initiatives in the public and private sector. Initiatives include an academic, medical center-based stress reduction program and a range of professional training programs and corporate workshops, courses, and retreats. There are links to the Stress Reduction Program; annual conferences; professional education; workplace programs; consultation and staff training; the Power of Mindfulness retreat; special programs on mediation and/or stress reduction; research to "deepen our understanding of mindfulness and its potential effects on mind and body, health and well-being;" and a bibliography of readings on meditation, mindfulness, and stress reduction.

Deep Psychology: The Quiet Way to Wisdom

Available: http://www.iawfonline.org/summit [Note: Click on link for 2005 8th Wildland Fire Safety Summit Proceedings, and follow the link for the presentation titled *Deep Psychology*.]

This paper is an executive summary of concepts presented in a slide show presentation by Ted Putnam at the 8th Annual Wildland Fire Safety Summit in Missoula, Montana (April 2005). Putnam reviews the eastern notion of the mind, and makes suggestions on how to improve mental functioning in everyday environments, especially under high risk environments. He puts forth meditation as "the single most powerful way to improve your mind and thus reduce accidents and fatalities". He recommends meditation training and practices to learn how to take one's mind off autopilot and "become more aware, see subtle interactions, analyze correctly and think clearer". Putnam's discussion concludes with a list of suggested readings for mindfulness meditation and the underlying psychology, as well as links to related websites. An expanded paper on this topic is being developed by Putnam.

Mind & Life Institute

Available: http://www.mindandlife.org

This non-profit organization fosters dialogue and research between modern science and Buddhism. The organization recognizes these as two ways of advancing knowledge, and it facilitates *experimental* and *experiential* science of the mind that can guide and inform medicine, neuroscience, psychology, education, and human development. Activities include meetings between prominent scientists and leading figures, such as the Dalai Lama; public conferences; publications based on conferences and other collaborative exchanges; the design of research between laboratory scientists, scholars and practitioners of Buddhism; and educational programs based on their research findings. The site links to a description of the interface between Buddhism and modern science, conferences and events, books and publications, audio and video media titles, research initiatives, and the Mind & Life Research Network.

The Science and Clinical Applications of Meditation

Available: http://www.investigatingthemind.org

This site reports on a 2005 conference titled "The Science and Clinical Applications of Meditation". The conference was sponsored by the Mind & Life Institute, Johns Hopkins School of Medicine, and Georgetown University Medical Center. It describes how meditation is becoming mainstream in western medicine and society, and describes the Dalai Lama's interest in, and dialogue with, western science. The site provides conference session descriptions, presentation abstracts, descriptions of speakers and panelists, and a link for ordering conference audio dvds and video cds.

CITATION INDEX

The following list of citations is intended to help readers find references to specific articles, books, book chapters, or reports that are included in this bibliography. The citations in this index are listed alphabetically, by the name of the first author. For articles that are cited on multiple pages in this document, an * marks the page containing the complete annotation.

KEYWORD INDEX

U

uncertainty 17, 20, 27, 40, 48, 52, 53, 56

V

values 14, 22, 26, 27, 30, 31, 35, 49, 65

W

worldview 21

Appendix A

Author's Picks: A Scaled-Down List of Suggested Readings

We recognize that this document contains a large number of readings and that working through them can be overwhelming, especially for the busy professional. Each of the annotated books, articles, reports, and case studies have been carefully chosen because of the insights they provide. However, to help readers gain familiarity in each topic area more quickly, we have included a scaled-down list of readings. We prepared this list based on our own favorites, as well as those of long-time members of the fire community who are knowledgeable about the topics and literature included in this document and have tried to integrate many of these concepts into fire operations. The following combination of readings would be, in our opinion, an excellent place to start one's reading and study. These readings are shown as "Author's Picks" throughout the document.

Decision Making and Sensemaking

Klein, Gary. 2000. **Sources of power: how people make decisions.** Cambridge, MA: MIT Press. 338 p.

Montgomery, Henry; Lipshitz, Raanan; Brehmer, Berndt, eds. 2004. **How professionals make decisions.** Mahwah, NJ: Lawrence Erlbaum Associates Publishers. 472 p.

Nutt, Paul C. 1999. **Surprising but true: half the decisions in organizations fail.** Academy of Management Executive. 13(4): 75-89.

Weick, Karl E. 2001b. **Making sense of the organization.** Malden, MA: Blackwell Publishers, Inc. 483 p.

Organizational Culture

Deal, Terrence E.; Kennedy, Allan A. 2000. **Corporate cultures: the rites and rituals of corporate life.** Cambridge, MA: Perseus Books Group. 232 p. [Reprinted from 1982.]

Schein, Edgar H. 1999. **The corporate culture survival guide.** San Francisco: Jossey-Bass Publishers. 224 p.

Thackaberry, Jennifer. 2004. **"Discursive opening" and closing in organizational self study: Culture as trap and tool in wildland firefighting safety.** Management Communication Quarterly. 17(3): 319-339.

Identification and Identity

Ashforth, Blake E. and Mael, Fred. 1989. **Social identity theory and the organization.** Academy of Management Review. 14(1) 20-39.

Bullis, Connie A.; Tompkins, Phillip K. 1989. **The forest ranger revisited: a study of control practices and identification.** Communication Monographs. 56: 287-306.

Gioia, Dennis; Thomas, James. 1996. **Identity, image, and issue interpretation: sensemaking during strategic change in academia.** Administrative Science Quarterly. 41(3): 370-403.

Leadership and Change

Deetz, Stanley A.; Tracy, Sarah J.; Simpson, Jennifer L. 2000. **Leading organizations through transition.** Thousand Oaks, CA: Sage. 232 p.

Heifetz, Ronald A. 1994. **Leadership without easy answers.** London: Belknap Press. 366 p.

Herzberg, Frederick. 2003. **One more time: how do you motivate employees?** Harvard Business Review. 81(1): 86-96. [Reprinted from 1968.]

Kotter, John P. 1996. **Leading change.** Boston, MA: Harvard Business School Press. 187 p.

Organizational Learning

Argyris, Chris. 1990. **Overcoming organizational defenses: facilitating organizational learning.** Needham Heights, MA: Allyn and Bacon. 180 p.

Garvin, David A. 2000. **Learning in action: a guide to putting the learning organization to work.** Boston: Harvard Business School Press. 272 p.

Senge, Peter M. 1990. **The fifth discipline.** New York: Doubleday. 432 p.

Senge, Peter; Kleiner, Art; Roberts, Charlotte; Ross, Richard; Smith, Bryan. 1994. **The fifth discipline fieldbook: strategies and tools for building a learning organization.** New York: Doubleday/Currency. 594 p.

Team and Crew Dynamics

Barker, James R. 1993. Tightening the iron cage: concertive control in self-managing teams. Administrative Science Quarterly. 38(3): 408-437.

Driessen, Jon. 2002. Crew cohesion, wildland fire transition, and fatalities. Missoula, MT: United States Department of Agriculture, Forest Service, Technology and Development Program, TE02P16—Fire and Aviation Management Technical Services. 15 p. Available: http://fsweb.mtdc.wo.fs.fed.us. [July 6, 2007].

Katzenbach, Jon R.; Smith, Douglas K. 1993. The wisdom of teams: creating the high-performance organization. New York: Harper Business. 352 p.

Scholtes, Peter R.; Joiner, Brian L.; Streibel, Barbara J. 2003. The team handbook. 3rd ed. Joiner/Oriel Inc. 400 p. Spiral bound.

Risk and Uncertainty

Kahneman, Daniel; Slovic, Paul; Tversky, Amos, eds. 1986. Judgment under uncertainty: heuristics and biases. New York: Cambridge University Press. 544 p.

Perrow, Charles. 1999. Normal accidents: living with high-risk technologies. Princeton, NJ: Princeton University Press. 386 p.

Reason, James. 1997. Managing the risks of organizational accidents. Brookfield, VT: Ashgate. 252 p.

Tompkins, Phillip K. 2005. Apollo, Challenger, Columbia: the decline of the space program. Los Angeles: Roxbury. 288 p.

Vaughan, Diane. 1996. The Challenger launch decision: risky technology, culture, and deviance at NASA. Chicago, IL: The University of Chicago Press. 592 p.

High Reliability Organizations

Keller, Paul, technical writer-editor. 2004. Managing the unexpected in prescribed fire and fire use operations: a workshop on the high reliability organization. Santa Fe, New Mexico, May 10-13, 2004. Gen. Tech. Rep. RMRS-GTR-137. Fort Collins, CO: U.S. Department of Agriculture, Forest Service, Rocky Mountain Research Station. 73 p. Available: http://www.wildfirelessons.net/HRO.aspx [July 6, 2007].

Roberts, Karlene. H., ed. 1993. New challenges to understanding organizations. New York: Macmillan. 256 p.

Roberts, Karlene H. 1990b. Managing high reliability organizations. California Management Review. 32(4): 101-113.

Weick, Karl; Sutcliffe, Kathleen. 2007. Managing the unexpected: resilient performance in an age of uncertainty. 2nd ed. San Francisco, CA: Jossey-Bass. 224 p.

Crisis Communication

Arvai, J.; Wilson, R.; Rivers, L.; Froschauer, A. 2004. Environmental risk communication: a review and annotated bibliography. Unpublished report. Eugene, OR: Decision Research. 165 p. [Note: This publication is available at the USDA Forest Service's Pacific Southwest Research Station via email (pwinter@fs.fed.us) or phone (951-680-1557).

Coombs, W. Timothy. 1999. Ongoing crisis communication: Planning, managing, and responding. Thousand Oaks, CA: Sage. 192 p.

Fink, Steven. 2002. Crisis management: planning for the inevitable. Lincoln, NE: iUniverse, Inc. 264 p.

Seeger, Matthew W.; Ulmer, Robert R. 2003. Explaining Enron: communication and responsible leadership. Management Communication Quarterly. 17(1): 58-84.